Build the TRS Drawbot.

www.radioshack.com/DIT

PARTS

- ○ Standard Hobby Servo 273-766
- ○ Small Clipboard
- ○ Metal Standoffs 276-195
- ○ Aluminum Offset Angle
- ○ Aluminum Flat Bar
- ○ Enclosed 4 x AA Battery Holder 270-409

- ○ Size M Panel Mount Power Jack 274-1563
- ○ I/8" Panel Mount Stereo Phone Jack 274-249
- ○ SuperLock Fasteners 64-2363/64-0258
- ○ Cable Clips 278-1640
- ○ Rubber Feet 64-2346/64-0256

TOOLS

- ○ Drill and Drill Bits
- ○ Hacksaw
- ○ Soldering Iron and Solder
- ○ Flat-Head Screwdriver
- ○ Phillips Screwdriver

4

Panel-mount the power jack in the ground rail, right over the battery pack, and solder a red jumper wire to its center-pin contact. Connect the other end of this wire to the power rail with a ring tongue. Cover the battery pack leads with heat-shrink tubing and install a power plug on their ends. Load batteries into the pack and plug it in to the power jack.

5

Cut a 6-I/2" length of I" aluminum flat bar and make perpendicular witness marks at I", 3-I/4" and 3-3/4" from one end. Mount a circular servo horn at one end as in Step I. Download the forearm template and bend the arm at the witness marks to match the profile. Attach two peel 'n' stick cable clips to hold the pen.

6

Download the calibration sound files. Turn on the battery pack, connect the headphone jack to the audio out on your laptop or other device and play the calibration sound. The servos will rotate to their default 90-degree positions. While the sound is playing, install the forearm on the elbow servo at a right angle to the upper arm and secure with the servo shaft screw. Repeat the process to install the shoulder servo on the base at a 45-degree angle.

7

Your TRS Drawbot is ready to go! Test it with the file square.wav. When you first play the file, the arm will move to its starting position and wait IO seconds for you to extend the pen tip. Then it will begin to draw! Depending on how accurately you were able to align the arms in Step 6, your square may be skewed slightly up or down on the page. Visit the project online for fine-tuning instructions, more sound files, and instructions on generating your own art.

CONTENTS

COLUMNS

22

FEATURES

SPECIAL SECTION
Robots

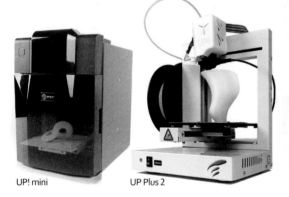

SKILL BUILDERS

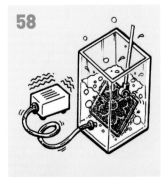

PROJECTS

TOOLBOX

Vol. 39, May 2014. MAKE (ISSN 1556-2336) is published bimonthly by Maker Media, Inc. in the months of January, March, May, July, September, and November. Maker Media is located at 1005 Gravenstein Hwy. North, Sebastopol, CA 95472, (707) 827-7000. SUBSCRIPTIONS: Send all subscription requests to MAKE, P.O. Box 17046, North Hollywood, CA 91615-9588 or subscribe online at makezine.com/offer or via phone at (866) 289-8847 (U.S. and Canada); all other countries call (818) 487-2037. Subscriptions are available for $34.95 for 1 year (6 issues) in the United States; in Canada: $39.95 USD; all other countries: $49.95 USD. Periodicals Postage Paid at Sebastopol, CA, and at additional mailing offices. POSTMASTER: Send address changes to MAKE, P.O. Box 17046, North Hollywood, CA 91615-9588. Canada Post Publications Mail Agreement Number 41129568. CANADA POSTMASTER: Send address changes to: Maker Media, PO Box 456, Niagara Falls, ON L2E 6V2

To show off
your High School's
innovative STEM program,
just bring it to a football game!

Eric Andracke is in his eighth year teaching high school Construction, Manufacturing, and 3D Animation courses. To share his school's commitment to STEM education, Mr. Andracke and his students brought their ShopBot Buddy and several class projects out to a football game. He explained, "It's so important for the community to see what is taking place inside of our classrooms. We figured the crowd was already gathered, and people were really excited to see our demonstration."

"STEM education," said Mr. Andracke, "bridges the gap between traditional math and science classes with hands-on activities, collaboration, and directive lessons."

Next year, Mr. Andracke's students will be building electric guitars, taking them from digital design through fabrication with the help of the ShopBot. The classwork will cover math calculations, science principles and equations, design, part engineering, and digital fabrication.

Make:

FOUNDER & CEO
Dale Dougherty
dale@makezine.com

CFO
Todd Sotkiewicz
todd@makermedia.com

> "The factory of the future will have only two employees, a man and a dog. The man will be there to feed the dog. The dog will be there to keep the man from touching the equipment."
> —Warren G. Bennis

EDITOR-IN-CHIEF
Mark Frauenfelder
markf@makezine.com

EDITORIAL

EXECUTIVE EDITOR
Mike Senese
msenese@makezine.com

MANAGING EDITOR
Cindy Lum

PROJECTS EDITOR
Keith Hammond
khammond@makezine.com

SENIOR EDITOR
Goli Mohammadi
goli@makezine.com

TECHNICAL EDITORS
Sean Michael Ragan
sragan@makezine.com
David Scheltema

DIGITAL FABRICATION EDITOR
Anna Kaziunas France

EDITORS
Laura Cochrane
Nathan Hurst

EDITORIAL ASSISTANT
Craig Couden

COPY EDITOR
Laurie Barton

PUBLISHER, BOOKS
Brian Jepson

EDITOR, BOOKS
Patrick DiJusto

CONTRIBUTING EDITORS
William Gurstelle, Nick Normal,
Charles Platt, Matt Richardson

CONTRIBUTING WRITERS
Greg Brandeau, Gareth Branwyn,
Eric Chu, Tyler Cooper, Eliot K.
Daughtry, Steve Davee, Jerome
Demers, Sam DeRose, Eva
Edleson, Max Edleson, Jeffrey M.
Goller, Nathan Goller-Deitsch,
Gregory Hayes, Steve Hoefer,
Jordan Husney, Grant Imahara,
Jon Kalish, Bob Knetzger, Russ
Martin, Gordon McComb,
Forrest Mims III, Dug North,
Windell Oskay, John Edgar Park,
Rick Schertle, Phil Shapiro,
Casey Shea, Naghi Sotoudeh,
Becky Stern, Keith Violette, Eric
Weinhoffer, Natalie Weirsma,
Tyler Winegarner, Sasha Wright

CONTRIBUTING ARTISTS
Matthew Billington, Jared
Haworth, Bob Knetzger,
Samantha Lucy, Greg Maxson,
Mark Rosenthal, Damien Scogin,
Natalie Stultz, Julie West

CREATIVE DIRECTOR
Jason Babler
jbabler@makezine.com

DESIGN, PHOTOGRAPHY & VIDEO

ART DIRECTOR
Juliann Brown

SENIOR DESIGNER
Pete Ivey

DESIGNER
Jim Burke

PHOTO EDITOR
Jeffrey Braverman

PHOTOGRAPHER
Gunther Kirsch

MULTIMEDIA PRODUCER
Emmanuel Mota

VIDEOGRAPHER
Nat Wilson-Heckathorn

FABRICATOR
Daniel Spangler

WEBSITE

WEB DEVELOPERS
Jake Spurlock
jspurlock@makezine.com
Cole Geissinger

WEB PRODUCERS
Bill Olson
David Beauchamp

ONLINE CONTRIBUTORS
Alasdair Allan, John Baichtal,
Jimmy DiResta, Haley Pierson-
Cox, Andrew Salomone,
Andrew Terranova

ENGINEERING INTERNS
Paloma Fautley, Sam Freeman,
Andrew Katz (jr.), Brian Melani,
Nick Parks, Sam Sheiner,
Wynter Woods

Comments may be sent to:
editor@makezine.com

Visit us online:
makezine.com

Follow us on Twitter:
@make @makerfaire
@craft @makershed

On Google+:
google.com/+make

On Facebook:
makemagazine

VICE PRESIDENT
Sherry Huss
sherry@makezine.com

SALES & ADVERTISING

SENIOR SALES MANAGER
Katie D. Kunde
katie@makezine.com

SALES MANAGERS
Cecily Benzon
cbenzon@makezine.com
Brigitte Kunde
brigitte@makezine.com

CLIENT SERVICES MANAGERS
Mara Lincoln
Miranda Mager

MARKETING COORDINATOR
Karlee Vincent

COMMERCE

DIRECTOR OF SHED DESIGN
Riley Wilkinson

OPERATIONS MANAGER
Rob Bullington

SENIOR CHANNEL MANAGER
Ilana Budanitsky

PRODUCT INNOVATION MANAGER
Michael Castor

MARKETING

VICE PRESIDENT OF MARKETING
Vickie Welch
vwelch@makezine.com

MARKETING PROGRAMS MANAGER
Suzanne Huston

MARKETING SERVICES COORDINATOR
Meg Mason

MARKETING RELATIONS COORDINATOR
Courtney Lentz

DIRECTOR, RETAIL MARKETING & OPERATIONS
Heather Harmon Cochran
heatherh@makezine.com

MAKER FAIRE

PRODUCER
Louise Glasgow

PROGRAM DIRECTOR
Sabrina Merlo

MARKETING & PR
Bridgette Vanderlaan

CUSTOM PROGRAMS

DIRECTOR
Michelle Hlubinka

CUSTOMER SERVICE

CUSTOMER CARE TEAM LEADER
Daniel Randolph
cs@readerservices.
makezine.com

Manage your account
online, including change
of address:
makezine.com/account
866-289-8847 toll-free
in U.S. and Canada
818-487-2037,
5 a.m.–5 p.m., PST
makezine.com

PUBLISHED BY

MAKER MEDIA, INC.
Dale Dougherty, CEO

Copyright © 2014
Maker Media, Inc.
All rights reserved.
Reproduction without
permission is prohibited.
Printed in the USA by
Schumann Printers, Inc.

CONTRIBUTORS

What was your most spectacular failure?

Samantha Lucy
Calgary, Alberta
(Steve Davee column illustration)
My biggest failure was attempting to cast a torso in chocolate. I learned that conceptual sculpture wasn't my thing and that chocolate mixed with wax can really ruin a crock pot.

Eliot K. Daughtry
Oakland, Calif.
(Robot taxonomist)
In sixth grade, I made a robot costume for myself out of a 30lb fiberboard barrel, dryer hose, window screen, and cardboard boxes, and was covered in a shiny metal backed with paper. It debuted in my neighborhood's 4th of July parade, in Richmond, Va., and I wore it for a 3-mile walk in sweltering heat, as it became heavier and heavier.

Eva Edleson
Deadwood, Ore.
(Wood-Fired Barrel Oven)
One time I had an earthen oven collapse while I was in the process of making it in a workshop! I didn't let this get me down, rebuilt the dome in a day, and still managed to get the project done in time. It all worked out and more people got to learn!

Gordon McComb
San Diego, Calif.
(Smooth Moods Sound Synthesizer)
One robot I'd sooner forget was a 200-pound "home assistant bot" that was too wide to fit through doorways. Lesson learned: Use a tape measure before you start building.

Steve Davee
Berkeley, Calif.
(Every Child a Maker)
With fond memories of a seven-story treehouse I built the summer I was 11, I naively figured as an adult I could build a 20-foot tower for $150 in three weeks. After three summers, student help, and $1,500, it stalled at 90% finished. Finishing details are tough.

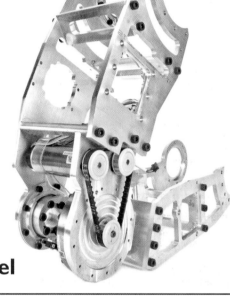

**WRITTEN BY
DALE DOUGHERTY,**
founder and CEO of
Maker Media.

Lessons Learned

CALL HIM THE QUINTESSENTIAL MAKER.
He's an accomplished builder who has ribbons from Maker Faire. He made his own website featuring a line of electronic kits he developed, along with tutorials. He has taught a four-hour intro class on Arduino to kids and adults at a hackerspace. He even spoke at an Atmel industry panel in New York City. Last October, he was invited to Maker Faire Rome, where he gave a talk about starting his own business. He met the mayor of Rome. Yet Quin Etnyre is just 13 years old.

Quin started reading MAKE magazine when his parents got him a subscription. Then he came to his first Maker Faire, where he learned to solder. The next year he came back to Maker Faire to exhibit his own project. Quin's mother says that when he was 10, he was getting bored with school and needed something more challenging. Quin found a new adventure through making.

Today anyone can design and develop a product. The barriers are lower and the technology is getting easier. It means that new people can become makers — and a self-directed 13-year-old can learn to do it. Quin is special, it's true. He has very supportive parents who give him access to the resources he needs, but he loves making.

"Going to school and running a business is hard," Quin says. "It is a lot of work." Quin's website, QTechKnow.com, is where his kits are listed. One of them is a fart detector, a kit designed for an audience that Quin understands better than most. He's learning about marketing and that his

customers really like to get stickers and pins with an order. Packing kits in bags and getting them to the post office takes a lot of time, but he says, "You have to be thinking of your next project." Quin also connects to other makers who serve as his mentors, many of whom do not live in his community.

I recently had a call from Jim Hogeboom, superintendent of the school district where Quin attends middle school. Quin had attended a school board meeting and told them what he was doing. "I made the point that I learned all of that outside of school," Quin tells me. "Schools should be teaching things like that during the school day because it has opened up so many cool opportunities for me." Quin is the kind of 21st-century learner that our schools aspire to produce, yet he's trying to decide what kind of high school he should attend next year.

Quin isn't just thinking about himself. He believes, like I do, that if schools engage more students as makers, the schools will be better and more students will discover new capabilities. Quin shared four ideas with the school board: "1) Offer more computer science courses, and higher-level courses for the geeks; 2) Put a hackerspace in every school with a great teacher who loves creativity and innovation; 3) Make this hackerspace open to all students in the school (not just the geeks) since makers can be useful in all fields of study — agriculture, music, art, psychology, etc. — and students interested in different fields can come together and collaborate on projects; and 4)

Make sure we start making at the elementary and middle schools, too."

Students like Quin are the agents of change in education. Mr. Hogeboom wanted to learn more about makerspaces in schools and how to develop the capacity for teachers to engage students as makers.

At the university level, Dr. Craig Forest is trying to provide more opportunities for students to make at the largest engineering school in the United States. Dr. Forest is an associate professor at the George W. Woodruff School of Mechanical Engineering at Georgia Tech in Atlanta, and he was named Engineering Educator of the Year in Georgia last year. He developed a student-centered makerspace called the Invention Studio (inventionstudio.gatech.edu) on campus.

The Invention Studio has become a place where any student could hang out, have access to tools, and do their own projects. The Invention Studio is open to personal projects as well as course or research projects. "From Halloween costumes to battlebots," says Dr. Forest, "students are learning how to design and build things they are passionate about and taking ownership of their education."

The most interesting feature of the makerspace is who runs it — the students. The Makers Club consists of 100 students who play a variety of roles in managing the space. "It's their space," states Dr. Forest. "And that's key." It's the Makers Club's responsibility to keep the place up and running, even welcoming newcomers and showing them how to use the equipment safely and responsibly. The Makers Club serves the needs of 1,000 users per week in the 24/7, free-to-use facility.

There are students who do well in the Invention Studio but don't always do well in their courses. A young woman had a 2.0 GPA in her courses, but she got a great job at a large company because of the projects and leadership skills she had developed in the Invention Studio.

With the experience students gain from designing and building things, and solving real problems, I hope they can find jobs that they really enjoy. Or perhaps they will consider starting their own ventures. If so, they should connect with Quin, whose talk at Maker Faire Rome was "Lessons Learned from a 12-Year-Old CEO." ◐

Gunther Kirsch

Maker Faire ®

YEAR OF THE MAKER

CELEBRATING COMMUNITY AROUND THE WORLD

GREATEST SHOW & TELL ON EARTH

BAY AREA
KANSAS CITY
DETROIT
NEW YORK
OSLO
UK
PARIS
ROME
TOKYO
SHENZHEN

5th Annual NEW YORK SEPT 20+21
makerfaire.com Brought to you by MAKE magazine

Drones, Cat Trackers, and Guitars

>> My son Colton (13 years old) has a subscription to your magazine. He loved your last edition with the license plate electric guitar [Volume 37, page 76]. He actually made it for a Destination Imagination project. Thank you for the great, inspiring magazine that you publish! Love seeing my son drop his video games to put his creativity to work!
—*Trina Page, Cross Roads, Texas*

>> I loved the article by William Grassie on "Quadcopter Photogrammetry" [Volume 37, page 42]. I'm a budding quadcopter enthusiast and a geomorphologist interested in finding new ways of building 3D models of the Earth's surface. I now have an idea for using quadcopter photogrammetry for collecting scientific data.
—*Dr. Rob Parker, Cardiff, Wales, UK*

>> My son Dylan and I have absolutely no experience with drones (with the exception of seeing drones flying around the slopestyle course during the Sochi Olympics), electronics, and/or soldering techniques and have never even held an R/C transmitter. Why we thought we could take on building one from scratch, I have no idea. Regardless, we undertook the task ["The HandyCopter UAV," Volume 37, page 44]. The experience was fantastic. The instructions supplied within MAKE were excellent. That, combined with the additional information supplied on the website, made the activity with my son exciting, challenging, and rewarding.
—*Stephen Shragg, Encino, Calif.*

>> I am so excited to finally be able to read my MAKE issues on my iPad. I have already used the app to show MAKE magazine and explain Maker Faire to an acquaintance, and I've used it for a parts reference while shopping at the local RadioShack. Thanks!
—*Mike Oitzman, Penn Valley, Calif.*

>> I just finished assembling one of the projects from MAKE: the GPS Cat Tracker [Volume 37, page 96]. It was a pretty sweet project, and I'm sure I'll be able to find a plethora of different uses for the finished product. Additionally, learning about the TinyDuino was worth the project just by itself. I'm totally struck by how insanely small it is, and I'm sure this won't be the last time I make use of one. I have one problem I'm running into now: The GPS is logging a *.txt* file to the microSD with no trace of an *.nmea*. Do you have a lead on to why this might be happening? Thanks in advance, and I think you've come up with a radical project here.
—*Zack Telfer, Jackson, Tenn.*

AUTHOR KEN BURNS RESPONDS:
>> The reason for the *.txt* is that the SD card library doesn't support a four-character file extension. However changing it to *gps.nmea* on your computer should allow it to be read fine. Thanks! ✪

MAKE AMENDS:

■ In Volume 37's "All Art Is Made by Makers" (page 10), it was Dutch master painter Johannes Vermeer who used camera obscura, not Edgar Degas.

■ In Volume 38's "Top Photography Hacks" (page 50), we failed to credit the discovery of the "turkey pan beauty dish" to creator Megan Abshire. Great job, Megan!

■ In Volume 38's "Maker Friendly Hardware Stores" (page 24), McGuckin's store manager is Randy Dilkes, not Randy Barker. Our apologies, Randy!

■ In Volume 38's Toolbox (page 102), the images for the USB-powered and battery-powered versions of the V.I.O. Stream Camera were erroneously swapped.

What Sticks About Play and Bricks

Every child is a natural-born maker. Written by Steve Davee ■ Illustration by Samantha Lucy

THE YOUNGEST SLOWLY TODDLED OVER TO HER OLDEST SISTER, the equivalent of a quarter of her body weight precariously clutched in her tiny fingers. With a grin, she dropped the muddy brick, narrowly missing their toes. "Oh! Thanks!" the 6-year-old said, tactfully ignoring the near miss, and that the brick came from a previously built section. "Maybe you can find a little bit smaller one, from that pile?" Their parents and I, observing from a distance, exchanged knowing nods. She's got this.

The oldest resumed, setting bricks in increasingly precarious stacks. The toddler hefted over a half brick, shuffled her feet out wide, and triumphantly threw it down between them. "But ... that doesn't fit with the others ..." Then a light went off. "Perfect!" she said. She placed the brick on top of a wobbly stack and added a full brick overlapping an adjacent stack, stabilizing both. She turned to rebuild in this stronger, staggered pattern, a reinvention of ancient masonry techniques.

Their middle sister was engrossed in scraping bricks — with the help of her stuffed animal — creating a fine, red dust. "Is that glue for the blocks?" the oldest asked. "Actually, it's fairy dust. And also glue. My dog is helping. He's getting dirty!" She began making designs on the bricks with a small pointed rock. "It's so pretty!" they said in unison.

A trio of brothers arrived, drawn in by the scene. The youngest, a new walker, stumbled over and enthusiastically toppled part of the wall. Seeing disappointment, his 5-year-old brother stepped in. "We can help you build a door there, then you don't have to step over walls to get inside!" All agreed and began playing together,

engineering an entrance, and decorating with mud and brick dust. "Now we need to name our house!"

I'm struck by the lessons of this episode and what I observed — tests of strength, confrontation of and adaptation to danger, collaboration, engineering and design iteration, curiosity, empathy, imagination, social finesse, and the pure joy of building, telling stories, and playing together. It

exemplified what we know making to be, all catalyzed by a pile of bricks in the muddy grass of a restaurant's backyard.

This, and countless episodes of play occurring every day, serves as a beautiful illustration that children are natural-born makers. Imagination, curiosity, playfulness, and compassion are among their intrinsic traits. These are just a few of the superpowers of children.

The Maker Education Initiative (makered. org) has a vision: "Every Child a Maker."

It's both a recognition of these childhood powers — that every child is born a maker in the purest, broadest, and most inclusive sense of the word — and an expression of the need to continually nurture and allow for these characteristics to flourish. All children deserve more opportunities to express themselves in a myriad of forms, to make in a million ways, and learn about the world and each other using their built-in tools of learning: play, imagination, and their constant drive to create.

What can we do to recognize and learn from these superpowers? How might education take advantage of these incredible traits and the myriad of languages children use to express creativity?

The Maker Movement and Maker Faires are helping to inspire educational change. We hear it every year, at every Faire: "Why can't schools be more like Maker Faire?" As MAKE founder Dale Dougherty has said, "At the heart of Maker Faire is this idea of play — we kind of get lost in it." As educators and caregivers strive to foster a modern resurgence of hands-on, project-, problem-, inquiry-, and play-based learning, we have the Maker Movement and the incredible capabilities of children to guide us. If we listen. If we allow the space, time, and places for play, exploration, and making to occur. We have much to learn from children, and so much to do to ensure that every child has more opportunities and possibilities in their lives. We need more piles of muddy bricks. ◗

STEVE DAVEE is the director of education and communications for the Maker Education Initiative (makered.org). He is a former biochemist, math and science teacher, documentation specialist, and the founder of CoLab Tinkering.

How Maker Faire Shaped My Future

From hobbies to Harvey Mudd via a salvaged aircraft. Written by Sam DeRose

Tony DeRose

EVERY SATURDAY MORNING FOR EIGHT MONTHS, I rolled out of bed into our garage to join the rest of the team. We were working on our most ambitious project to date: The Viper, a flight simulator that we designed, built, and fundraised ourselves. On weekdays we were high-school students, but on the weekends we built a spinning, two-axis motion gantry out of a salvaged airplane fuselage, created custom software around an open-source flight simulator game, and managed a website with hundreds of photos and dozens of update videos, attracting widespread internet attention. Thanks to donations from Nvidia , Autodesk, and a Kickstarter campaign that raised $16,000, we were able to debut The Viper at Maker Faire 2012, where we shared our project with thousands of attendees.

However, this story starts way before high school. It was on my 12th birthday that I got tickets to the very first Maker Faire, back in 2006. The battling bots and giant flaming sculptures captured my attention so much that I requested to go back the next year. The third year, my dad and I got ambitious and decided to exhibit a project: a 2'x3' multitouch display for a computer.

As a 14-year-old, I got hooked on explaining and demonstrating our exhibit to people who were as interested and excited about our project as we were. After two days of discussing the physics of light refraction and the best ways to diffuse LEDs with complete strangers, I realized how special Maker Faire is. People who make really cool stuff in their garages congregate from all over, and the resulting atmosphere is an addictive blend of wacky new ideas, collaboration, and shared excitement. From then on, it became an annual tradition to create something for Maker Faire.

As our projects broadened to include numerous different areas of making, so did our interests, skills, and connections with other makers in the area. As a result, the creations we exhibited at Maker Faire became increasingly elaborate. It took five years of this snowball effect for us to work up to The Viper.

> **"People who make really cool stuff in their garages congregate from all over, and the resulting atmosphere is an addictive blend of wacky new ideas, collaboration, and shared excitement."**

For the eight months of the build, I was completely enveloped in The Viper. Even though we only worked as a team on weekends, most of us spent the rest of the week mulling over clever design solutions, torque calculations, and how we were possibly going to finish in time for the Faire. I constantly got asked why I was slaving away on something that didn't really do anything important. The answer seemed obvious to me, and when we finally finished the Viper and brought it to Maker Faire, it was obvious to everyone at the Faire as well. For us makers, it seems like a simple answer: Because it's awesome, that's why!

But the real reason lies deeper. It's a truly rewarding experience creating something that inspires and excites people, or that causes them to stop, think, and ask a question. Why did I devote months of my life to The Viper? Because it was projects like The Viper that originally inspired me back when I was 12 at the first Maker Faire.

Making started as a hobby, but has turned into more. When I was applying to colleges, I zeroed in on Harvey Mudd because it has a fantastic machine shop and a student body who uses it. At Harvey Mudd, I feel the same shared excitement as I do at Maker Faire, and for the first time in an academic setting, I'm surrounded by peers who think and make like I do. Every job or summer internship I've had has been based off one question: How much cool stuff do I get to make?

I've realized I want to spend my life working on exciting projects with a team of people who are as obsessed with making as I am. This realization is not only responsible for how I spend my free time, it's currently shaping my career path. So get inspired, make stuff, repeat. ⊘

SAM DEROSE
loves woodworking, cooking, electronics, sewing, and just about anything that involves creating fun new stuff. He's currently a sophomore at Harvey Mudd in Los Angeles, but his true home is the San Francisco Bay Area.

MADE ON EARTH

#madeonearth

The world of backyard technology

Know a project that would be perfect for Made on Earth?
Email us: *editor@makezine.com*

PAPER PERFECT

ERICSTANDLEY.30ART.COM

Virginia-based fine artist **Eric Standley** is busy blowing minds with his laser-cut paper sculptures reminiscent of impossibly detailed stained-glass windows. Standley spends up to six months drawing and planning each piece, determining depth and obsessively playing with dimension and details.

An associate professor of studio art at Virginia Tech, Standley conceptualized the series by chance while working on a project where he was cutting Cheerios boxes. He decided to try cutting them intricately with the laser cutter and "more or less to be absurd," he recalls. When he saw the cut sheets stacked atop one another, he found himself drawn in. "To be conscious of drawing on those multiple layers at one time and think about a whole composition with depth, I knew there was something interesting here — I was using a different part of my brain to draw."

Standley creates vector drawings for each layer before beginning the laser-cutting process, which can take upwards of 60 hours, considering that many of his sculptures use more than 100 sheets of paper. Virtually a paper architect, he must take into account structural elements, implementing curves and overlaps in the design to stabilize unsupported arches. Lest folks think technology is making his task easier, Standley adds, "Every efficiency that I gain through technology, the void is immediately filled with the question, 'Can I make it more complex?'"

— *Goli Mohammadi*

Eric Standley

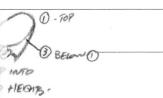

① - TOP
③ BELOW ①
INTO
HEIGHTS.

THROUGH-FLOAT
CENTERED TO
SMALL CIRCLE.

PROFILE CIRCLE
INTO SIDE @
.75" WALL
FOR THE
THROUGH FLOAT

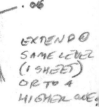

.02
.06

EXTEND @
SAME LEVEL
(1 SHEETS)
OR TO A
HIGHER ONE.

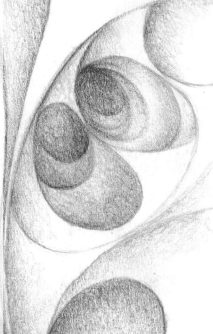

IS THAT A TRAIN CAR IN YOUR BASEMENT?

KINGSTONSUB.COM

Canadian train enthusiast **Jason Shron** has a special place in his heart for Via trains, Canada's intercity passenger rail cars. When he was 12, he wrote Via a letter asking if he could buy a seat from one of the trains, but the return letter said no.

Fast forward to his adulthood, when he heard that Via coach #5647 was to be scrapped. With a group of friends, he was able to salvage it. More than four and a half years and 2,500 hours of build time later, Shron has reconstructed the train car in a 12′×20′ room in his basement complete with all the authentic bells and whistles.

The garbage can, coat hooks, '70s carpet, radiator covers, folding gate, first aid box, and comfy chairs are all there. He replaced the bathroom with his record collection and sound system, and even added a photomural of the next car that's visible through the train doors, truly creating the illusion of being on a real train.

To others hoping to follow his lead, he offers, "You, too, can have a train in your basement. You need two things: to be completely insane and to have an amazing wife."

— *Goli Mohammadi*

Jason Shron

BUZZ BOOTH

MAKEZINE.COM/BEEBOOTH

Bees are clearly the makers of the insect world. And just like humans, some bees prefer working alone to working amongst the distracting buzz of social collaboration. Luckily, Toronto artist **Sarah Peebles** and a team of collaborators have created Audio Bee Booths, where native solitary bees (not honeybees) and wasps can take care of their industrious insect business.

Consisting of a wood nesting plank with different sized grooves routed into it, the booths are designed to accommodate a diverse array of tunneling bee and wasp species. The cutaway side of the nesting plank is covered with a piece of Plexiglas to give visitors a close-up look at the bees' domestic lives. The booths also feature vibrational sensors embedded in the nesting planks that pick up the sounds the bees and wasps make and amplify them through headphones.

The result is that visitors to the pyrography-adorned Audio Bee Booths are afforded the uniquely intimate experience of hearing and seeing everything that the booths' reclusive inhabitants are up to.
— *Andrew Salomone*

✚ Learn to make your own solitary bee condos: makezine.com/solitary-bee-condos

Sarah Peebles

Robert Cruickshank

This Audio Bee Booth installation is by Sarah Peebles, assisted by Rob Cruickshank, electronics; John Kuisma, woodworking; and Chris Bennett, pyrography.

HOMEMADE CLACKULATOR

SIMONWINDER.COM/PROJECTS/RELAY-CALCULATING-ENGINE

When you think "pocket calculator," your next thought probably isn't "custom cabinetry." But for **Simon Winder** in Seattle, his single-operation relay calculating engine, built over five years, with over eight months of full-time work, deserved nothing less. The cabinet was built by local artisan **Matthew Richter**.

"You can easily see how these devices [electromechanical relays] are being used, in contrast to modern computers where the operations happen far removed from human senses," muses Winder.

The machine has one function: to calculate and display a square root. The user enters a number on a rotary dial, whose pulses are stored in binary by the system of relays. The start button activates a mechanical clock pulse generator that triggers every operation state with precision timing.

A full calculation takes about a minute. While the operations cascade through 480 relay switches, hand-wired into circuits, each closing relay declares itself by lighting up with a satisfying mechanical clack. A bell announces completion of the final calculation, and the answer displays on eight nixie tubes.

Winder will be bringing his fantastic contraption to Maker Faire Bay Area 2014.

— *Gregory Hayes*

Ed Shively

REACH FOR THE STARS SCOTT.J38.NET/INTERACTIVE/REACH

As part of the Tough Art residency at the Children's Museum of Pittsburgh, designer **Scott Garner** created Reach, a large-scale interactive mural and musical instrument. When someone touches both the moon and a star, a tone is played. A "triumphant chord" is the result when all of the stars are touched simultaneously.

"Pittsburgh is a city of bridges, so the idea of spanning or reaching across a space came directly from the skyline," says Garner. "I was also interested in the idea of noncompetitive play, in which a child could accomplish a goal alone, or by cooperating with their family and friends."

The panels of the installation are constructed of Baltic birch plywood painted in a textured metallic style. To make the aluminum stars, Garner used the water jet cutter at Pittsburgh's TechShop. "I spent a lot of time filing them down to make them safer for little fingers."

As for the final result, "After a few tentative taps, kids are often sprawled across the piece, touching stars with all limbs and giggling."

— *Matt Richardson*

Kristi Jan Hoover

100 MAKER FAIRES
around the world

IN 2013, MAKERS CONVENED IN 20 COUNTRIES TO SHARE, LEARN, BUILD, AND INNOVATE.

NINE YEARS AGO, WHEN WE HOSTED OUR VERY FIRST MAKER FAIRE IN SAN MATEO, CALIF., little did we know that "the Greatest Show (and Tell) on Earth" would become the physical epicenter of a global phenomenon coined the Maker Movement. In 2013, there were 100 Maker Faires across the globe, hosted in diverse locations ranging from big cities like Rome, Tokyo, and New York City to small towns like Loveland, Colorado, and Machynlleth, Wales. Notably, 93 of these Faires were smaller-scale, locally produced "Mini Maker Faires," up from just 56 in 2012.

These Faires are bringing makers together, encouraging collaboration, innovation, sharing of knowledge, and a celebration of all things maker-made. To boot, Maker Faires are providing a viable showcase for local startups, maker businesses, and new products, enabling makers to pursue their passions and make a living doing what they love. We're seeing more and more products launched and new companies announced at the Faires.

Among last year's Faire organizers, 36 were museums, 20 were nonprofits, and 12 were makerspaces. We've seen organizers go on to influence local education initiatives and affect change in their school districts. With Faires that are in their subsequent years, we're hearing how local makerspaces have grown as a result, new ones have been started, and the local maker community has been mobilized. Emily Smith, who spearheaded the Vancouver Mini Maker Faire, now in its fourth year, shares, "Hosting a Maker Faire meant creating a new community that didn't otherwise exist, by bringing together many groups."

In 2014, there are 140 Faires projected, and as founder Dale Dougherty said, it looks to be the Year of the Maker.

MAP
(AS OF APRIL, 2014)
- Flagship Faires
- Featured Faires
- Current Minis
- Mini Maker Faire Applicants
- Past Events

Anchorage Mini Maker Faire

Vancouver Mini Maker Faire

Maker Faire Kansas City

Maker Faire Detroit

World Maker Faire New York

Maker Faire Bay Area

Oaxaca Mini Maker Faire

Bogotá Mini Maker Faire

Santiago Mini Maker Faire

IN 2013:
530,000 ATTENDEES
55 FIRST-YEAR EVENTS
36 MUSEUM-ORGANIZED MINI MAKER FAIRES
91% WOULD ATTEND AGAIN
1,700,000 ATTENDEES IN 8 YEARS

Maker Faire Bay Area
San Mateo, Calif.

Maker Faire Kansas City
Kansas City, Mo.

Maker Faire Detroit
Detroit, Mich.

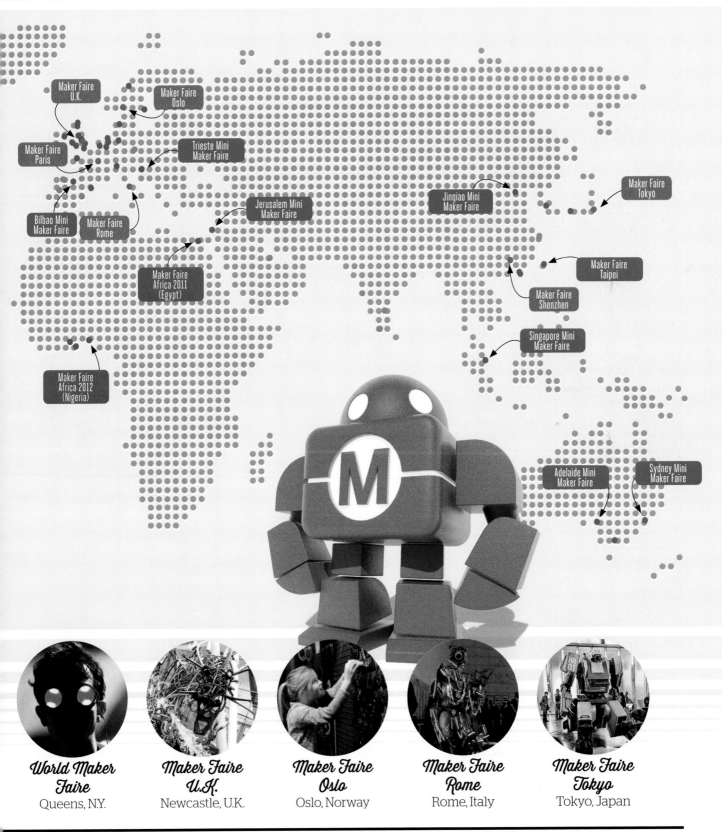

Maker Faire
U.K.

Maker Faire
Oslo

Maker Faire
Paris

Trieste Mini
Maker Faire

Maker Faire
Tokyo

Jinqiao Mini
Maker Faire

Bilbao Mini
Maker Faire

Maker Faire
Rome

Jerusalem Mini
Maker Faire

Maker Faire
Taipei

Maker Faire
Africa 2011
(Egypt)

Maker Faire
Shenzhen

Singapore Mini
Maker Faire

Maker Faire
Africa 2012
(Nigeria)

Adelaide Mini
Maker Faire

Sydney Mini
Maker Faire

World Maker Faire
Queens, N.Y.

Maker Faire U.K.
Newcastle, U.K.

Maker Faire Oslo
Oslo, Norway

Maker Faire Rome
Rome, Italy

Maker Faire Tokyo
Tokyo, Japan

MAKER FAIRE CALENDAR
Mid-May through July 2014

- **MAKER FAIRE BAY AREA (flagship)**
  (San Mateo, Calif.): May 17 & 18

- Stockholm Mini Maker Faire
  (Sweden): May 17 & 18

- Trieste Mini Maker Faire
  (Italy): May 17

- Mendocino County Mini
  Maker Faire (Calif.): May 24

- Maker Faire Taipei
  (Taiwan): May 24 & 25

- Torino Mini Maker Faire
  (Italy): May 31

- Jerusalem Mini Maker Faire
  (Israel): June 5—7

- Eugene Mini Maker Faire
  (Ore.): June 6

- Maker Faire North Carolina
  (Raleigh, N.C.): June 7

- Reno Mini Maker Faire
  (Nev.): June 7

- Montreal Mini Maker Faire
  (Canada): June 7 & 8

- Vancouver Mini Maker Faire
  (Canada): June 7 & 8

- Columbia Mini Maker Faire
  (S.C.): June 14

- Waterloo Mini Maker Faire
  (Canada): June 14

- **MAKER FAIRE PARIS (featured)**
  (France): June 21 & 22

- McAllen Mini Maker Faire
  (Texas): June 21

- Barcelona Mini Maker Faire
  (Spain): June 22

- **MAKER FAIRE KANSAS CITY (featured)**
  (Mo.): June 28 & 29

- Maker Faire Hannover
  (Germany): July 5 & 6

- Bilbao Mini Maker Faire
  (Spain) July 12 & 13

- Kingsport Mini Maker Faire
  (Tenn.): July 13

- SolarFest Mini Maker Faire
  (Vt.): July 19 & 20

- Anchorage Mini Maker Faire
  (Alaska): July 26

- Singapore Mini Maker Faire
  (Singapore): July 26 & 27

- **MAKER FAIRE DETROIT (featured)**
  (Mich.): July 26 & 27

- Manchester Mini Maker Faire
  (U.K.): July 26 & 27

COMMUNITY-BASED, INDEPENDENTLY PRODUCED MAKER FAIRES ARE TAKING PLACE ALL OVER THE GLOBE. EACH BEARS THE UNIQUE FLAVOR OF ITS LOCALE, AS SHOWN IN THESE POSTERS. FIND A FAIRE NEAR YOU OR LEARN HOW TO ORGANIZE YOUR OWN AT **makerfaire.com/map**.

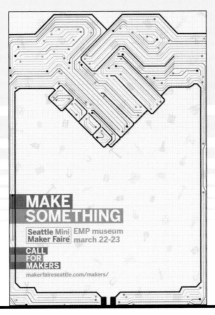

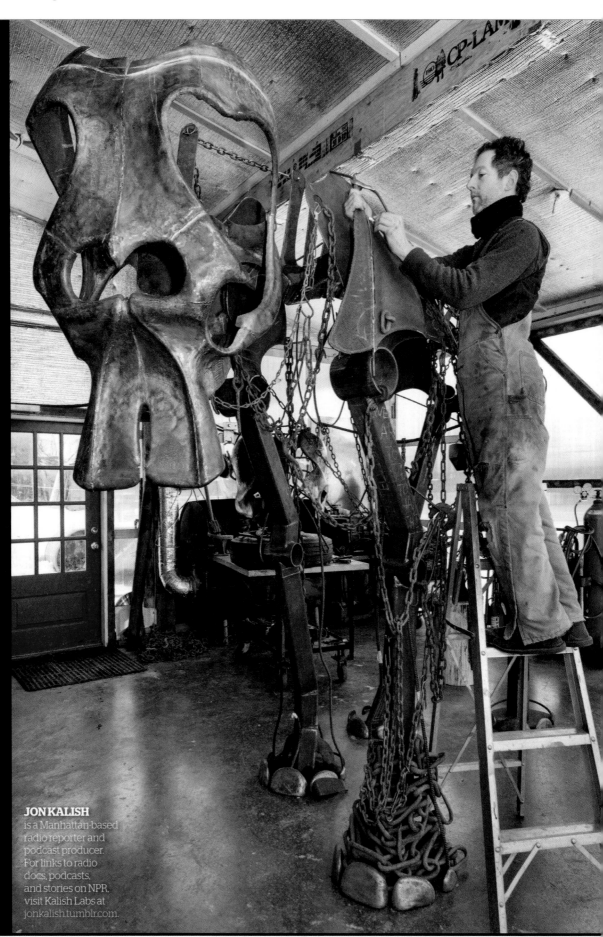

THE SELF-TAUGHT SCULPTOR

Written by Jon Kalish

JON KALISH is a Manhattan-based radio reporter and podcast producer. For links to radio docs, podcasts, and stories on NPR, visit Kalish Labs at jonkalish.tumblr.com.

EBEN MARKOWSKI'S METALWORKING SKILLS ARE ROOTED IN HIS FATHER'S CAR RESTORATION SHOP.

EBEN MARKOWSKI IS NOT A STARVING ARTIST. That's because he's dedicated to reducing overhead and living life in its simplest terms. While Eben has previously managed to cobble together a living by doing stints of excavator work and restoring fancy Italian sports cars at his family's auto shop, the 38-year-old Vermont sculptor, who makes nearly life-sized renditions of animals, would like to do nothing but make art. He just may get to the point where he can do that thanks to an Asian elephant and its calf.

"We live paycheck to paycheck," Eben says, referring to his wife, Heidi, a chef and woodworker, and many members of the animal kingdom who live on their nearly 13-acre homestead in Panton, Vt., about a mile from Lake Champlain.

You'd think that two Vermonters who are not exactly rolling in dough would be reluctant to spend several thousand dollars a year feeding and providing medical care for enough critters to fill Noah's Ark. Yet the Markowski spread has ducks, chickens, sheep, goats, a cow, dogs, and a cat. The cow, Petunia, was brought to the Markowskis by "a veterinarian with a heart" when a local farmer had left her for dead. Other adoptees, including an old ram named Chester, came from Shelburne Farms, an environmental education center north of this dairy farm area.

Eben's connection to animals goes way back. "You could look at my mother's photo albums and the earliest photos show me holding a frog or a crayfish or a worm," he says.

His ability to create realistic sculptures of animals also dates back to his childhood. His father, Peter, remembers Eben at the tender age of three announcing one day, "I made a rabbit." When his father asked to see it, little Eben opened his hand and revealed what his father remembers as "an unbelievable, small, to-scale rabbit" made out of Play-Doh.

Eben makes a very good hourly wage working for his father, something akin to what a master plumber earns. But the sculptor is trying to minimize his time in the family auto shop, Restoration and Performance Motorcars, because of his commitment to sustainable living. Eben dismisses the shop's focus on rebuilding old Ferraris and other exotic sports cars as "dinosaur work. We should really be doubling down on what would sustain us."

And yet Eben acknowledges that the family business has played a pivotal role in his career as a sculptor. It was inside the two burgundy-colored barns his father built for auto restoration in Vergennes, Vt., that Eben had the tools and workspace to sculpt. And it was at the family business that his first patrons stumbled upon his works in progress.

"My father's car shop was a tremendously fortunate place to be," he tells me as we walk past a 1954 Ferrari 750 Monza, an Aston Martin from the late 1930s, and a spotless Ford Woody station wagon from the 1940s.

Eben and his younger brother Stephan were in the auto shop at a very young age. Eben was about 10 when he was taught to weld. Having a father who fixed things all the time was a factor. "You don't throw something away, you fix it," he was taught as a boy.

> **"You could look at my mother's photo albums and the earliest photos show me holding a frog or a crayfish or a worm."**

Jerry Swope

Today Eben jokes that he and his brother were indentured servants when they were boys but then quickly stresses that they loved working on cars. There were times when their father didn't let customers know just how young the shop's "mechanics" were.

"My father would put the phone to his chest and say, 'Guys, come on.' And he'd shoo us out from under cars because a client was coming in and he didn't want the client to see two little kids putting in a drive shaft," Eben chuckles.

On one occasion a client of the car restoration business became a client of the sculptor. When Eben was 20 years old, he was about halfway through completing a giraffe sculpture when a man and his wife stopped in.

"I remember them driving by and abruptly pulling in and asking what it was that I was doing," the sculptor recalls. "Fortunately for me, they saw the potential in it. One of the most flattering things for an artist is when someone wants something that you're making. They can see where you're going and you have inspired them to want it."

Before he finished the first giraffe, another couple rolled in to the shop and commissioned a second giraffe. At about 18 feet tall, it turned out to be bigger than the first one.

Eben also did a five-panel rhinoceros relief made with sheets of copper roofing hammered on wooden forms he made using a handheld grinder with a blade attached. It measures 6 feet by 12 feet and is 5 inches deep.

While Eben was working on the rhino, he took on a challenging car project: He spent a year and a half making an entire new body for a priceless 1951 Ferrari, working from a handful of half-century-old grainy photographs of the vehicle and employing traditional techniques used in the '50s. His artistic aspirations dovetailed nicely with the needs of his dad's business and created a unique niche for him.

"Instead of having to say, 'Oh, we need to have this door panel made. We need to go to Boston or California to find someone to do sheet metal,' I happened to be there teaching myself how to do this for sculptures," Eben recalls.

Eben "not only understands animal forms but mechanical forms and machines and movements," observes Homer Wells, a painter and tinkerer who lives in Monkton, Vt. "Eben's powers of observation are absolutely unbelievable, and his mechanical voodoo is beyond belief," Wells says.

Those skills allow him to live what he

> **"Eben's powers of observation are absolutely unbelievable, and his mechanical voodoo is beyond belief."**

considers a more environmentally responsible life, one where frugality, doing-it-yourself, and recycling are central tenets. This is evident in his handmade home. "Eben welds and I'm a woodworker. The two of us together can make almost anything," Heidi says proudly.

It's also apparent in his tools and building materials. An unapologetic dumpster diver and self-confessed collector of scrap material, Eben modified a 1981 Ford backhoe that originally came with four different levers to operate its rear shovel. He cannibalized a discarded printing press for bearings and gears to make a linkage that enabled the shovel to be controlled with two levers instead.

A work in progress, his root cellar roof is actually half of an old fuel tank made of ¼-inch-thick steel. His youngest brother, Judd, cut the 16-foot-long tank with an oxyacetylene torch. The half tank weighs around 2,500 pounds and sits on the root cellar's stone walls, which are 4 feet high and as thick as 3 feet in some spots. Standing inside the root cellar, Eben looks up at the ceiling's orangey-brown rust patina and declares, "I think rust is totally underrated."

There are rebar hoops welded to the root cellar ceiling to hang garlic for storage. Heidi planted 9,000 heads to grow the crop commercially. A portion of their new 720-square-foot sculpture studio will be used for garlic drying.

Heidi helped her husband build the 24-foot by 30-foot studio onto the side of their house, which took about six weeks. It has 100-year-old hand-hewn hemlock timbers for framing and 10mm polycarbonate greenhouse glazing for walls, creating a solarium effect that makes the studio 25° warmer than outside.

The studio also has a 10-foot-high door, which Eben points out is big enough for an elephant. In winter of 2013, the sculptor went to work on his latest commission: a life-sized Asian elephant and its calf. He expects to complete the main elephant in early summer 2014 and then get started on the calf.

As for going back to work in the family car business, Eben says he won't rule that out if he's in a financial bind or his family really needs him for a project. "We try not to give him the mundane tasks because his head isn't in it," his father explains. "But when it's a huge challenge or something really off the wall, he's all about that." ⊘

+ See more of Eben's work at ebenmarkowski.com

Written by Charles Platt

Adventures in BODY COOLING

HOW A SELF-TAUGHT MAKER FOUND HIMSELF BUILDING MEDICAL PROTOTYPES.

CHARLES PLATT is the author of *Make: Electronics,* an introductory guide for all ages. He has completed a sequel, *Make: More Electronics,* and is also the author of Volume One of the *Encyclopedia of Electronic Components.* Volumes Two and Three are in preparation. makershed.com/platt

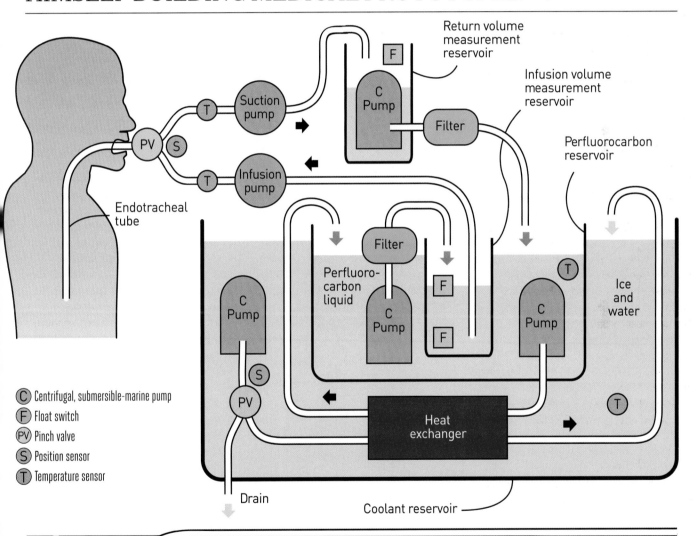

- Ⓒ Centrifugal, submersible-marine pump
- Ⓕ Float switch
- ⒫⒱ Pinch valve
- Ⓢ Position sensor
- Ⓣ Temperature sensor

IF SOMEONE SUFFERS A HEART ATTACK, THE RELATIVELY BRIEF INTERRUPTION IN BLOOD FLOW CAN LEAD TO PERMANENT BRAIN DAMAGE. The risk of this damage can be reduced by lowering the body temperature by 3°C to 5°C. Some hospitals try to achieve this by using cold gel packs, while others are equipped with immersion devices such as the "Thermosuit," which circulates ice-cold water around the patient.

Unfortunately, there's a snag. Cooling should be rapid, to maximize its therapeutic effects, but drawing heat out of the body is a slow, inefficient process. Even total immersion in ice and water requires an hour to lower the temperature by 5°C.

Various ideas have been proposed to accelerate the cooling process. The most radical would use liquid ventilation, which means infusing the lungs with a chilled, breathable liquid. Perfluo-

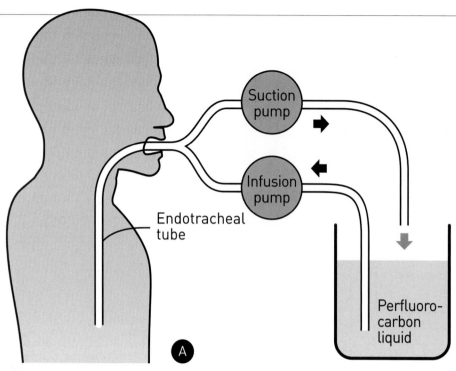

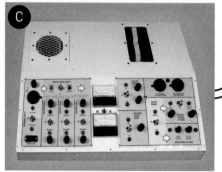

Suction pump

Infusion pump

Endotracheal tube

Perfluoro-carbon liquid

A

B

C

rocarbon liquids were proved to be breathable long ago in research sponsored by the Naval Research Laboratory, and a mixture of perfluorocarbon liquid and oxygen has been used to displace fluid from lungs in infants suffering acute respiratory distress syndrome.

In 2002, I wrote an article for *Discover* magazine about Critical Care Research, a small laboratory in California where Michael Darwin and Dr. Steven B. Harris were exploring liquid ventilation for rapid body cooling. The cold liquid would extract heat from blood in the network of capillaries throughout the lungs, and the blood would then cool the brain from within, far more efficiently than external cooling.

The procedure had not been perfected at that time, but I kept in touch with the staff at the laboratory — never guessing that I might become more personally involved.

By 2004, after perfecting a benchtop system, the lab hired an engineering company to make a portable, compact version, as proof-of-concept to demonstrate that something similar could be deployed by paramedics in an ambulance. Alas, the prototype didn't work very well, and a second engineering company didn't do much better. While chatting with the people at Critical Care, I said that it looked to me as if the engineers had made everything too complicated. Of course, I had no qualifications, but

I was arrogant enough to think that I could make it work by keeping it simple. Imagine my surprise when I was told to try.

Liquid ventilation basically requires one pump to infuse the lungs with liquid, and another to suction it out (**Figure A**). The difficult part is to make this happen precisely, reliably, and safely. Any prototype that I built would never be used on a human patient, but it would have to be theoretically usable, with proper control and monitoring of tem-

> **“I realized that I had made the same mistake as the engineering companies: I had been seduced by complexity.”**

perature, pressure, liquid volume, and cycle time. Because liquids tend to slosh around unpredictably, and because perfluorocarbon liquid is about twice as dense as water, with a very low surface tension that causes it to leak easily, building a prototype might be a bit more challenging than I had imagined.

Fortunately, the lungs are not sterile, so they don't require heavy-duty medical-grade pumps. I speculated that 12-volt centrifugal marine pumps could be sufficient, so long as there was minimal flow resistance.

Measuring the volume accurately was

the big challenge. One of the engineering companies had tried to weigh each infusion, but the rapid intermittent motion of the liquid made this difficult. I bought an overpriced, fancy-looking flow meter from a laboratory supply house, but the impeller inside it had a hopelessly nonlinear response. So, through eBay I ordered a British device for measuring draft beer dispensed in pubs. For purely financial reasons, this had to be accurate, but it created too much backpressure. I located other flow meters in cooling systems for overclocked computer CPUs; these were simple, accurate, and cheap, but they suffered a lag time when flow began, and they overran when it stopped. What to do?

One morning I woke up remembering that during a drought emergency in my youth, homeowners had been encouraged to place a brick in each toilet tank, to reduce the volume of water when the toilet was flushed. So here was the answer. An additional pump would fill a miniature tank. Its capacity would determine the infusion volume, and could be adjustable for experimental purposes by inserting little calibrated plastic tabs. A float switch would stop the infusion pump when the tank was empty, and a speed control on the infusion pump would adjust the infusion throughout a preset cycle time. This utterly basic system turned out to be accurate within +/–5ml.

My finished prototype (**Figure B**) placed the perfluorocarbon in a central reservoir, surrounded by a larger container of ice and water, which eliminated bulky insulation and a power-hungry refrigeration unit. My control panel used a few toggle switches and a pair of potentiometers (**Figure C**). Two 555 chips controlled the cycle times, and three NiMH battery packs delivered 25A for up to 45 minutes (**Figure D**). The whole thing had rather a homemade look, and I wondered if it would satisfy my clients. I transported it to California to find out.

To everyone's surprise, my portable version was actually more efficient than the lab-bench version. My clients were happy, so they did what happy clients sometimes do: They asked for more. I received an offer to move to California for a year, where I would set up my own workshop near the lab at company expense, to build a new, more sophisticated, more powerful prototype.

This was a very strange thing for a journalist to be doing. I found myself in a small industrial park, wrestling with pumps and tubing while a professional welder visited a warehouse area at the back to create components based on my plans. He was like a human 3D printer, using stainless steel instead of plastic. I was in fabricator heaven.

Within a year, we had a more compact, efficient, successfully tested prototype (**Figure E**), and I created a set of drawings in the primitive style required for a patent application (**Figure F**). But now there was a new request: Automate everything, with error messages, prompts, and safety interlocks.

This was much more difficult. First I tried using a tablet computer, but it was vulnerable to water, vibration, and input errors. I defaulted to pushbuttons and microcontrollers, but there were so many inputs and outputs, I needed seven controllers (two of them being 40-pin chips). Getting them to talk to each other reliably was a nightmare.

Eventually I had a functional circuit, but when I transplanted it from breadboard to perforated board, capacitive effects and voltage spikes emerged from nowhere. Worse, a microcontroller-compatible flash-memory chip turned out to have a truly horrible undocumented feature: its write speed diminished as the number of stored bytes increased. I didn't discover this until all the components had been soldered in.

I took a break to write a feature about privately funded space ventures ("Rocket Men," MAKE Volume 24, page 62). This introduced me to a remarkable man who had been the systems integrator at JPL for the first Mars Rover. When I showed him the plans for my liquid ventilation project, he said it looked like a climate control system for a space suit. It was "nontrivial," but he could fix it for $20,000, and it would only take him a couple of months. That sounded good to me, but my friends were getting impatient, and there was some reluctance about the extra expense. They decided they didn't need fully automated control for further testing, and I realized, with chagrin, that I had made the same mistake as the engineering companies: I had been seduced by complexity.

My final version used just one microcontroller but combined all the knowledge we had gained about the liquid ventilation process (**Figure G**). Now my clients were happy again — but I was suffering from fabrication burnout. Components, such as a specially graduated reservoir or a low-current, low-voltage pinch valve, had to be handmade with great precision. It was difficult, time-consuming — and ultimately quite boring.

Today, my friends at the lab are still doing research with my final version of the liquid ventilation equipment. It has a peak cooling rate of around 1° per minute, which is 10 times the claimed rate for cooling by immersion. We received a patent, and eventually, perhaps, an FDA-approved version will save thousands of lives per year by extending the window of opportunity for resuscitation after cardiac arrest. I think a large medical equipment company will have to build a human-rated system and take it through the regulatory process.

I was immensely privileged to participate in this adventure, and I'm proud of the outcome. Still, I'm happy to be back writing books. My current project, the *Encyclopedia of Electronic Components Volume Two*, is another task that has turned out to be more challenging than I expected — but still much easier than my adventures in cooling the human body. ◉

+ Special thanks to Saul Kent and Steven B. Harris, Joan O'Farrell, Sandra Russell, and Mike Lee of Critical Care Research.

D

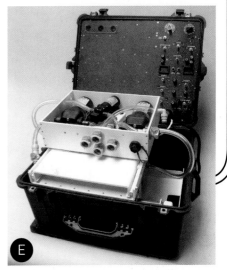

E

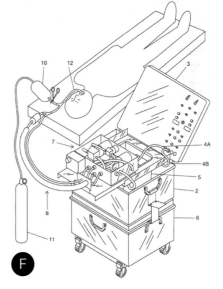

F

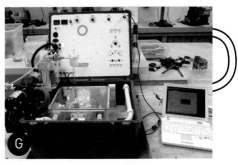

G

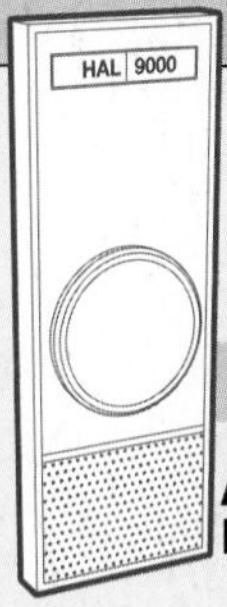

Artificial Intelligence

Computer

Disembodied Intelligence

PROGRAMS

Worker

[WE, ROBOTS']

Welcome to the factory.

FROM THE WORD'S FIRST USE IN KAREL ČAPEK'S 1920 play *R.U.R.*, "robots" have existed in fiction before fact. Čapek's *roboti* were not metallic, but biological "people" built by (and for) an assembly line. And though the science of robotics is so far mostly about electronics and mechanics, it may not always be. What will it be like when we *can* remake ourselves, for real? What will it mean to be manufactured?

Taxonomy by Eliot K. Daughtry ■ Illustrations by Greg Maxson

HUMANS

Transgenic

Mutant

CYBORGS

Synthetic

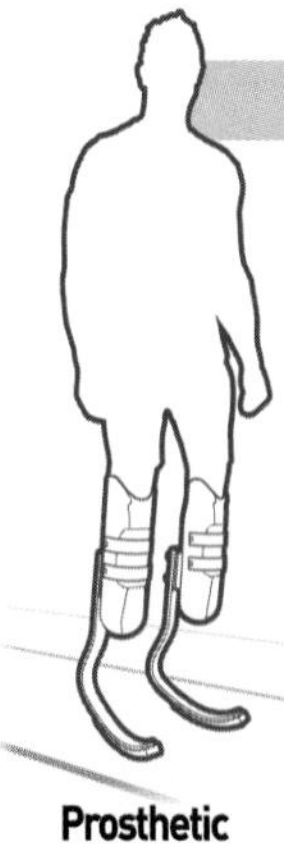

Prosthetic

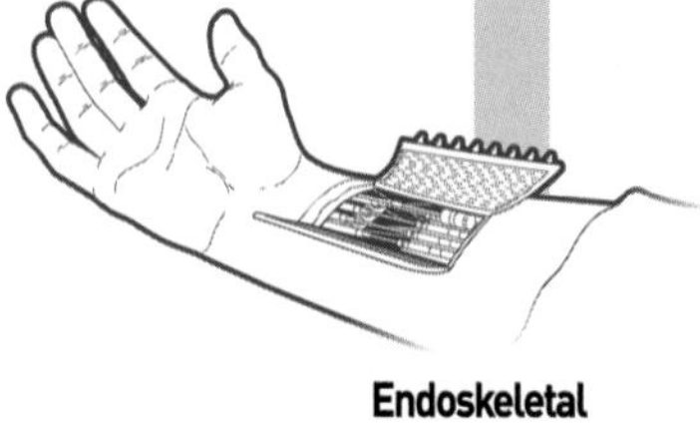

Endoskeletal

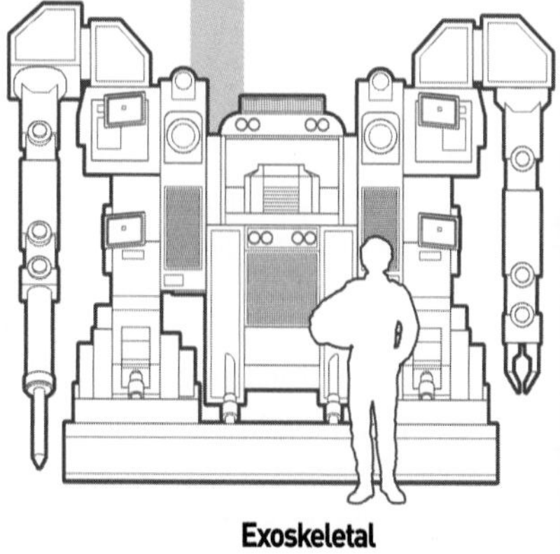

Exoskeletal

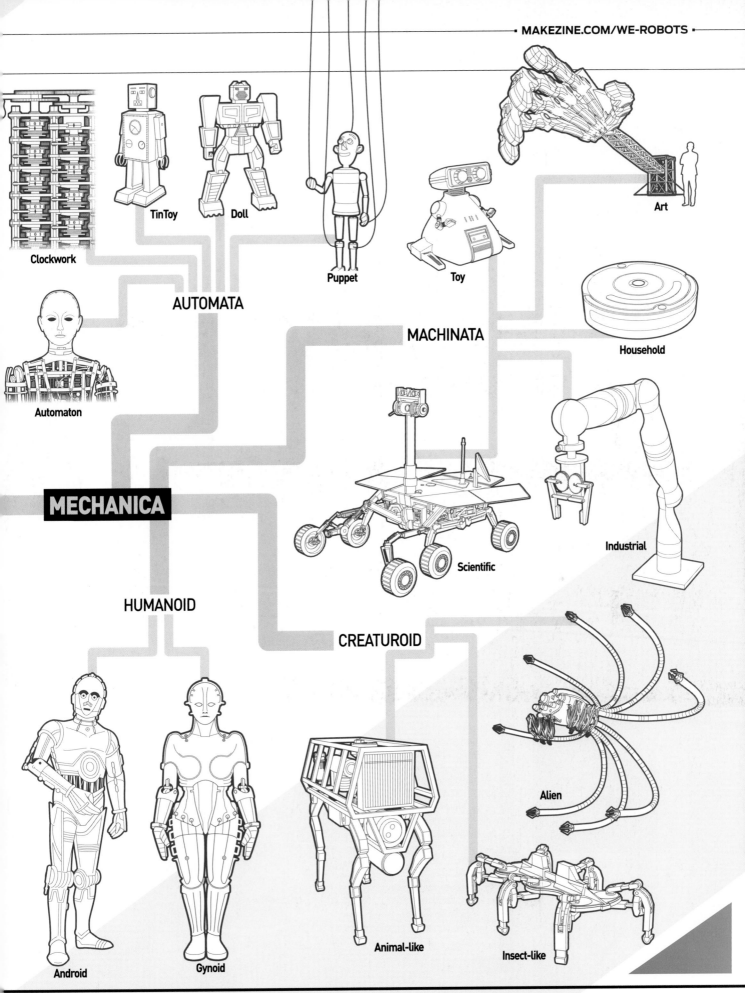

Clockwork

TinToy

Doll

Puppet

Toy

Art

AUTOMATA

MACHINATA

Household

Automaton

MECHANICA

Scientific

Industrial

HUMANOID

CREATUROID

Alien

Android

Gynoid

Animal-like

Insect-like

IN OUR OWN IMAGE

Written by Gareth Branwyn

The state of the art in humanoid robotics

IT'S THE PERENNIAL DEBATE: R2-D2 VERSUS C-3PO, Astromech versus Protocol Droid, utility versus usability, function versus form. When the maker movement took off 10 years ago, a bottom-up, keep-it-super-simple approach to amateur robotics made more sense, especially from a cost perspective. But now, thanks to cheaper, smaller control systems and access to sophisticated technologies like 3D printing and networkable servos, more makers are starting to tackle much more complex, fussier, high-end robotics projects like androids, gynoids, and anthropomorphic arms. Here are just a handful of the most exciting developments in the world of anthrobotics — whether military, commercial, or maker-made.

GARETH BRANWYN is a freelancer writer and the former Editorial Director of Maker Media. He is the author or editor of a dozen books on DIY tech and geek culture, including the first book about the web (*Mosaic Quick Tour*) and the *Absolute Beginner's Guide to Building Robots*. He is currently working on a best-of collection of his writing, called *Borg Like Me*.

ATLAS

This towering 6-foot biped is the latest from Boston Dynamics, makers of such frighteningly biomorphic bots as BigDog and Cheetah. With the introduction of Atlas, the Massachusetts-based military contractor lumbers a few steps closer to Terminator-style weaponized androids. Atlas has 28 hydraulic-actuated degrees of freedom, plus fully articulated hands to allow not just the lifting and carrying of objects, but the use of tools designed for humans. It can even climb with its hands and feet. Currently, Atlas is powered by a cable to an external supply. No word on when it's going to be upgraded to onboard power.

courtesy of Boston Dynamics

THE POPPY PLATFORM

This 33"-tall open-source humanoid platform is under development by Flowers Lab in Bordeaux, France. Designed for experiments in bipedal locomotion and other science, art, and education applications, Poppy uses strong, lightweight 3D-printed structural components. Twenty-five Robotis Dynamixel networkable servos are controlled by PyPot, a custom Python framework developed for the project. The unique inward bend of Poppy's hips (mimicking human hips) greatly improves the bot's walking efficiency. Given the high cost of Dynamixel servos, Poppy costs around $10,000 to build — but whoever said humanoid minions come cheap?

THE SHADOW DEXTEROUS HAND

Designed to mimic the human hand as closely as possible, the Shadow Dexterous has 20 actuated degrees of freedom and four actuated movements for a total of 24 joints. Each joint has a range of motion similar to that of a human hand. The system uses embedded PIC microcontrollers and PSoCs (Programmable Systems-on-Chips) in the hand itself for embedded control and pressure-sensing, plus EtherCAT protocol for overall control via a connected PC. At $120,000, you may have to black-market one of your biological hands to pay for one of these.

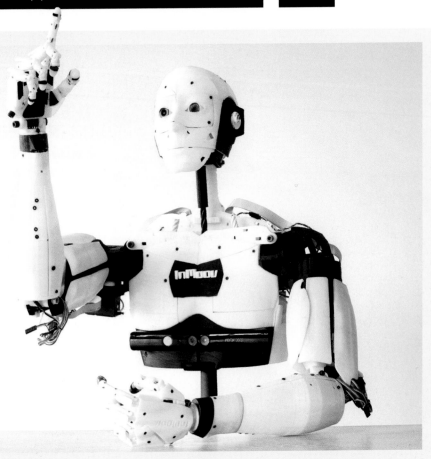

INMOOV

Launched in 2012 by French model maker and sculptor Gael Langevin, InMoov is an ambitious and inspiring project to create an open-source 3D-printable humanoid. On his site, Langevin hosts all the files you need to download and print your own mini-humanoid — at least (so far) the torso, arms, and head. The project has attracted an international community of builders and programmers intent on developing and refining the remaining parts of the robot.

IN OUR OWN IMAGE

THE JACO ARM

The JACO Lightweight Robotic Arm is a beautifully designed six-axis manipulator with a three-fingered hand. Created by Canadian company Kinova, JACO is named after co-founder Charles Deguire's uncle Jacques, who built robotic arms in the '80s to help overcome his own muscular dystrophy. The carbon fiber-housed arm offers six degrees of freedom and can handle a maximum payload of 1.5kg. The arm's API is accessible via USB 2.0. This level of sophistication comes at a price. The JACO sells for a wallet-squeezing $48,400.

THE OPEN HAND PROJECT

This volunteer-driven, open-source initiative works to develop a 3D-printable robotic hand that can be used either on robots or on human amputees. By crowd-sourcing development, using desktop 3D printing and other accessible maker technologies, indie U.K. robotics developer Joel Gibbard hopes to reduce the cost of a prosthetic hand from $100,000 to $1,000.

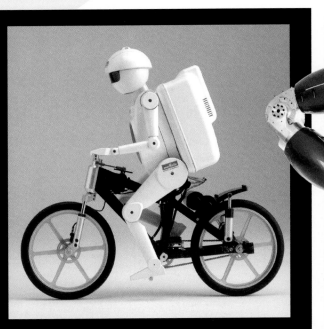

MURATA BOY

This 20" (50cm) bike-riding robot is a kind of mascot for Japanese electronics component company Murata. The doll-sized device can balance, pedal forwards and back, and avoid obstacles. Balance is achieved through three gyros, including one controlling a flywheel in the robot's chest. A Bluetooth-connected gestural wand directs the robot's movements. His younger cousin, Murata Girl, rides a unicycle.

ROLLIN' JUSTIN

This teleoperated humanoid/wheeled hybrid platform offers all the benefits of an anthropomorphic upper body without the challenges and hassles of bipedal stabilization and walking. The bot was developed by the German Aerospace Center (DLR) at The Robotics and Mechatronics Center in Bavaria. The "Justin" upper half was designed as a spacebot that can be remotely operated from Earth for applications like satellite repair and space station maintenance. The "Rollin'" configuration was intended for more down-to-Earth domestic chores.

TORO

DLR's Torque-Controlled Humanoid Robot is the product of lessons learned developing the anthropomorphic "Justin" torso, combined with bipedal legbots the German company has also been developing since 2009. TORO incorporates torque sensors in its leg joints to read and react to pressure and impact, giving a more robust and dynamic walking locomotion. TORO also has remarkably small feet (enabled by the advanced leg tech), allowing it to climb obstacles other robots can't manage.

For links, additional images, and more amazing anthrobotic art, please visit makezine.com/anthrobots
Share it: #anthrobots

DLR (CC-BY 3.0)

MY HOLLYWOOD

As a *MythBuster*, model maker, and combat roboticist,

Jeffrey Braverman

DREAM MACHINES

I've hacked everything from R2-D2 to the Energizer Bunny.

MY LOVE OF ENGINEERING GOES WAY BACK. When I was 3 years old, I got my first Lego set. I would say it changed my life, but at that age, I really didn't have much experience to go on, so better to say it helped shape the way I look at solving problems. I learned to work within a modular system and to use the parts at my disposal to build what I imagined. And what I imagined, essentially, was a career in engineering.

When I was in high school, there were no STEM (or STEAM) programs. There were no Lego Mindstorms kits or kids' robotics programs like FIRST. In fact, I only learned what an engineer was from a college guidance counselor. When she described it, a light clicked on in my head, and I said, "Yes, *that's* what I want to do."

Over the years I have blown up home theater equipment, helped create feature films, and used science to debunk urban legends. And while special effects, explosions, and car crashes are exciting, the thing I have enjoyed most is making robots. All kinds of robots. From mundane to downright dangerous.

Here are just a few old friends and favorites.

"THE SPIDER"

This monster started with the idea to create a large-scale walking robot strong enough to carry a person. I began in May 2007, with a six-legged design based on a Lynxmotion hexapod kit. I scaled up the joints and used a waterjet to cut the leg sections and body panels out of ⅝"-thick 6061 aluminum.

Each leg has a heavy-duty motor to lift it and a lighter-duty motor to move it forward and back. The leg design went through several revisions. I tried four different sets of motor/gearbox/external gear combinations before arriving at just the right speed and torque for the lifting movement. My external gear choices were limited due to space in and around each leg, and I eventually ended up using wheelchair motors with built-in electromagnetic brakes.

The control system is built around a Parallax BS2p40 microcontroller and PSC servo controller connected to an even dozen Vantec RBSA23 servo drivers. The microcontroller also runs a set of solid-state output modules that automatically engage and disengage the motor brakes so the robot won't waste power holding itself up while standing still.

Testing the Spider — certainly the heaviest robot I've ever created at more than 625lbs — proved to be a tremendous challenge. Working late at night by myself, there were a few too-close calls when the robot almost crushed me.

PRO TIP: Don't do what I did. Never work alone around heavy or otherwise dangerous equipment.

Control glitches also proved costly. I went through several $100 cast-iron gears when the legs bottomed out against their stops and the motor kept trying to turn, resulting in gear teeth shearing off and raining down on the floor like Tic Tacs. What has kept the project on the shelf for the past several years is inadequate torque in the motors that swing the legs back and forth — unfortunately, the major part of the "walking" movement. They were selected for their compact size, but they will need to be replaced with larger, more powerful motors before the project can move (literally and figuratively) to the next step.

Written by Grant Imahara

KEY COMPONENTS
- » Parallax BS2p40 microcontroller
- » Parallax PSC servo controller
- » 6 × NPC B81/B82 motors
- » 6 × NPC EJ818 motors
- » 12 × Vantec RBSA23 ("Bully") servo amplifiers
- » 6 × ODC5 solid state output modules
- » 2 × NPC-B1812 12V 22Ah sealed lead-acid AGM batteries

FUN FACT
Moving the Spider into my garage at home was a touch-and-go process. Loading it onto the truck was done in minutes (thanks to the *MythBusters* forklift), but unloading it at home took 4 people almost 6 hours, and is one of the most dangerous non-work-related things I've ever done.

GRANT IMAHARA
As a co-host of Discovery Channel's *MythBusters*, Grant has swum with sharks, spun a merry-go-round with a shot from a sniper rifle, hung beneath an airborne helicopter, fired cheese from a cannon into the San Francisco Bay, let two dozen tarantulas tap-dance on his head, dropped a BMW from a Sikorsky, cut a car in half with a rocket sled, and built many awesome robots. All in the name of science.

Before *MythBusters*, Grant lit the lights in R2-D2's dome and gave the Energizer Bunny his beat as an animatronics engineer and model maker for Industrial Light & Magic. His work appears in several blockbusters including *The Matrix* sequels, *Galaxy Quest*, *Terminator 3*, and *Star Wars: Episodes I-III* (though he is in *no way* responsible for Jar Jar).

"DEADBLOW" (BATTLEBOTS)

At the 1995 Robot Wars in San Francisco, I saw an awesomely powerful, relentless hammer 'bot named "Thor" that inspired me to design something similar, but using pneumatic (air) power rather than hydraulics. My version uses a paintball nitrogen tank for compressed gas storage, an air accumulator, and a high-flow, high-speed valve to drive the air piston back and forth, swinging the hammer.

We were up against robots with intimidating names like "BioHazard" and "Vlad the Impaler," so I decided to name my hammer robot "Deadblow," after the tool of the same name. (A dead-blow hammer is one designed to minimize recoil after a strike.) I drew up a 3D CAD model, then laser-cut the prototype from acrylic. When I was satisfied with the fit of all the components and panels, I used a CNC mill to cut "final" parts in aluminum.

But then, competing with fighting robots is a life of constant upgrades. I ended up going through three or four armor configurations, six different drive systems, and tons of minor and major weapon-arm and air-system tweaks to squeeze more performance out of my parts.

People often ask whether it's worth it to put so much time and money into something that's likely to be destroyed. My response? BattleBots is like a really cool party where your robot is your ticket to enter. It's about testing your ideas against smart, tough competitors, and about the thrill of combat. Taking damage is part of the fun. And bringing home a giant nut (the trophy) isn't bad, either.

KEY COMPONENTS

» **6AL4V titanium armor and weapon arm**
» **2024 aluminum superstructure**
» **S7 tool steel hammer head**
» **NiMH battery packs**

FUN FACT

Deadblow was so popular that it spawned a line of official toys, and I was asked to write a book, which I called *Kickin' Bot: An Illustrated Guide to Building Combat Robots.*

"ARTOO"

The time: May, 1997. The place: Leavesden Studios, UK. Filming of *Star Wars: Episode I*, was about to begin, and things were not going smoothly on the Theed hangar set. Specifically, the R2 units were not going smoothly — they kept getting caught in the door track. My team at Industrial Light & Magic had scant weeks to get the aging prop robots in shape for production.

Up to then, all the R2 units had some form of caster wheel in the front foot pod. Our solution to the door problem was to use a larger-diameter wheel that wouldn't fall into the crack, and mount it on an axle that would be actively steered along with the rest of the robot. We used wheelchair motors for locomotion, which are quiet and precise. I handled the power electronics and the radio control system, including the mixing for the new steering components.

Before filming of *Star Wars: Episode II*, I was called on to update the electronics on the whole R2 fleet, starting with the dome lights (aka "logic displays"). Previously, big bundles of fiber optics terminated at a rotating color wheel illuminated by a bright halogen lamp (unchanged since the early '80s), which gave a swirling appearance. I replaced the halogen/color wheel combo with two hockey puck-sized LED arrays driven by a microcontroller running pseudo-random PWM sequencer code (originally developed for the warp engines on the Protector in *Galaxy Quest*). I also made up a custom PCB to combine all the lighting functions in one neat little package.

Nelson Hall

KEY COMPONENTS
» **2 × Invacare Power9000 wheelchair motors**
» **2 × Seiko Tonegawa SSPS-105 servos (dome and steered wheel)**
» **RC receiver**
» **Microchip PIC16C series 8-bit microcontroller**
» **Vantec RDFR33 electronic speed control**

FUN FACT
The R2 unit we made for *Episode I* eventually became the "hero" unit — the one most used in filming close-ups.

"THE ENERGIZER BUNNY"

In early 2008, the Eveready Battery Company commissioned Industrial Light & Magic to build a new generation of Energizer Bunny mascots for TV commercial production. My eight-person team had a three-month timeframe to complete three new mechanical bunnies plus two nonmotorized "posers." My responsibility: all of the electronics and radio control systems.

Each Bunny has a custom circuit board and custom RC relays, with multiple 8-bit microcontrollers for interpreting the radio control commands and executing multistep movements — for example: *stop beating*, *raise arms*, *twirl sticks*, *stop twirling*, *lower arms*, and *resume beating*.

The internal structure is aluminum, with a vacuum-formed styrene outer shell. In total, the Bunny has 10 servos, 3 accessory motors, and 2 drive motors, and needs three operators: one for the head, one for the arms, and one to drive.

One of the most challenging parts of the project was that (for legal, truth-in-advertising reasons) the Bunny had to actually run on consumer-grade Energizer batteries. But with more than a dozen onboard motors — including two heavy-duty drive motors — the Bunny had massive power requirements, far outside the capacity of off-the-shelf batteries and battery packs. My solution was to wire tons of them in parallel. I stuffed a total of 44 AA batteries into a banana-shaped pack, like an AK-47 clip, which hides along with all the other electronics in the drum body.

KEY COMPONENTS
»3 × RC receivers
»2 × Microchip PIC16C series 8-bit microcontrollers
»4 × RC speed controls
»10 × Futaba S9402 servos
»44 × AA batteries

FUN FACT:
The Bunnies have names! They are E, F, and G — aka Earl, Floyd, and Garth.

"GEOFF PETERSON"

In 2010, Craig Ferguson asked me to design and build a "robot skeleton sidekick" to help him host *The Late Late Show* on CBS. "Keep it simple," he told me.

I used a standard plastic biology classroom skeleton, but replaced the torso with an aluminum plate where I mounted the servos and other electronics. "Geoff" is equipped with three high-power servos — one for each arm and one in the neck — capable of delivering 27 foot-pounds of torque each. There's also a smaller hobby-type servo in the head, which flaps the jaw.

Originally, Geoff had a complex, iPad-based wireless control system and onboard wi-fi router. A microcontroller interpreted trigger commands sent from the iPad and played a corresponding sound file, as well as orchestrating a set of synchronized servo movements to give Geoff the illusion of life. Basically, you would push a button, and the robot would respond on its own. Before each show, the producers would load up a bunch of preprogrammed clips and then select from that library to respond to Craig. (Incidentally, this is how "balls" became one of Geoff's catchphrases, since it could be used in many different situations.) Lately, they have been operating Geoff with a live puppeteer, which allows greater spontaneity and many more comedic options.

Upgrades are always planned for Geoff, and recently, he got the ability to blink his eyes.

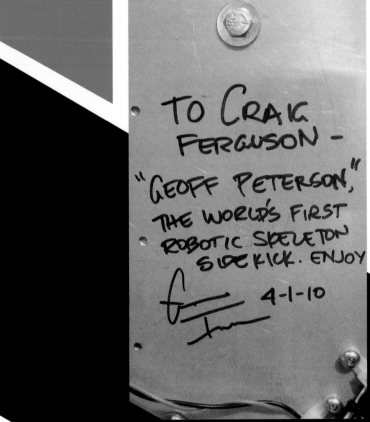

TO CRAIG FERGUSON —
"GEOFF PETERSON",
THE WORLD'S FIRST
ROBOTIC SKELETON
SIDEKICK. ENJOY
4-1-10

KEY COMPONENTS
›› **Parallax Basic Stamp microcontroller**
›› **Parallax Propeller servo controller**
›› **SparkFun MP3 Trigger**
›› **Wi-fi router**
›› **3 × Seiko Tonegawa SSPS-105 servos**
›› **Futaba S9402 servo**
›› **Wiznet Ethernet interface**
›› **2 × switching power supplies**

FUN FACT
Craig himself recorded the first version of Geoff's voice. He sounded like a Dalek with a Scottish accent. Then actor Josh Robert Thompson took over voicing and controlling Geoff and does so to this day.

"ODDJOB"

The weapon of choice for the evil henchman of James Bond's nemesis in *Goldfinger* is a razor-rimmed bowler hat he throws with deadly speed and accuracy. For the *MythBusters* test, we built our own bowler-*chakram* with a sharpened steel ring inside. Unfortunately, it turned out to be too heavy for a human to throw accurately.

We needed superhuman strength, speed, and precision, so I built a steel-framed robot with a throwing arm powered by a massive pneumatic cylinder. The arm had an "elbow," a second air cylinder to snap the "forearm," and a pneumatic "finger" at the end that held onto the hat. The cylinder was pressurized ahead of time and the arm was held under tension by a 1-ton quick-release latch. The trick was that the timing had to be extremely precise. I used limit switches to trigger the air valves; as the main arm swung 'round, it hit these switches, which caused the forearm to extend and the finger to release the hat at just the right moments.

KEY COMPONENTS
» **4.5" dia. × 24" stroke pneumatic cylinder**
» **2" dia. × 10" stroke pneumatic cylinder**
» **1.5" dia. × 6" stroke pneumatic cylinder**
» **2 × high-speed, high-volume pneumatic valves**
» **1-ton quick-release latch**
» **ANSI #80 sprocket and roller chain**
» **Limit switches**

FUN FACT
The steel-rimmed hat actually lopped the head off the target statue on the second try, and we were incredibly excited ... until we discovered that our "solid marble" statue was, in fact, hollow. Subsequently, we used a solid concrete statue, which the hat was only able to chip.

"THE BRIDE"

In *Kill Bill*, Uma Thurman's character The Bride is buried alive in a wooden casket. Rather than accept her fate, she uses her martial arts skills to punch the coffin lid over and over until it collapses, and then she digs her way out. For the *MythBusters* test, we needed someone who could punch repeatedly for hours — with phenomenal strength — without giving up. The list of human volunteers was surprisingly short, so we opted for a robot.

To build it, I used a small air tank as an accumulator and attached a large, high-speed/high-volume air valve as well as four smaller air valves. The main air valve handled the powerful punching stroke, while the smaller valves handled the retraction. We connected them to a squat (but powerful) air cylinder equipped with a fist cast from ballistic gelatin. The whole thing was connected to a giant external air supply and controlled by a pair of digital interval timers.

We fired the robot up, and it proceeded to go THUMP, THUMP, THUMP on the coffin lid for hours. After approximately 600 punches, it did actually create a crack in the wood, but it didn't punch through. Sorry, Uma! Rest in peace! ◉

For more photos from Grant's scrapbook, plus our video tour of his workshop, please visit makezine.com/imahara
Share it: *#imaharabots*

KEY COMPONENTS

» **Dual OMRON digital interval timers**
» **4 × small pneumatic valves**
» **High-flow/volume pneumatic valve**
» **4.5" dia. × 10" stroke pneumatic cylinder**

FUN FACT

The main high-speed/high-volume valve was borrowed from special effects company M5 Industries, and several smaller air valves came from my personal fighting robot, Deadblow.

ROBOT BUILDING BLOCKS

When I think about all my builds over the years, I find there are a few components I use over and over again — kind of like my own personal Lego set. Here are my Top 10, and why I reach for them again and again:

1. SHAFT KEYS: Yes, set screws can hold rotating gears or wheels on shafts, but they'll often slip under load. Shaft keys, on the other hand, *never* slip.

2. SHAFT COLLARS: If you use both shaft keys and shaft collars, your set screw could fall completely out and you'd still be fine.

3. MOUNTED BEARINGS: Bearings are essential in moving robots, but they require precise tolerances for mounting. After I discovered premounted bearings, I never looked back.

4. FLEXIBLE SHAFT COUPLINGS: Join distant components quickly and easily without having to know or care exactly how far apart they may be. As a bonus, they add a little bit of cushion.

5. HOSE CLAMPS: Great for securing small components to frames. No holes needed. Just wrap it around a tube and tighten with a screwdriver. Quick, easy, and adjustable.

6. BARRIER STRIPS: If the lifeblood of a robot is power, then the wiring is its circulatory system. Barrier strips make it easy to distributing that power as needed and allow you to easily reconfigure wiring without cutting and/or soldering.

7. VELCRO: Sometimes you just need a temporary connection. I use both the adhesive-backed "tape" variety (for mounting small components like radio control receivers and battery packs), and the two-sided "strap" kind (for cinching down large batteries).

8. CRIMP CONNECTORS: Make sure your connections stay connected, but that they can also be modified without too much trouble. A reliable crimping tool is absolutely essential.

9. SWITCHES: Use toggle switches to turn things on and off, and momentary switches for detecting when a mechanical component has reached a desired position.

10. CABLE TIES: Tame your wiring and keep it out of harm's way. *Not* to be used for structural applications — unless you're really desperate.

BUILD A REMOTE-CONTROL VENUS FLYTRAP

Explore biorobotics with a carnivorous plant that bites on command!

Written by Jordan Husney and Dr. Sasha Wright

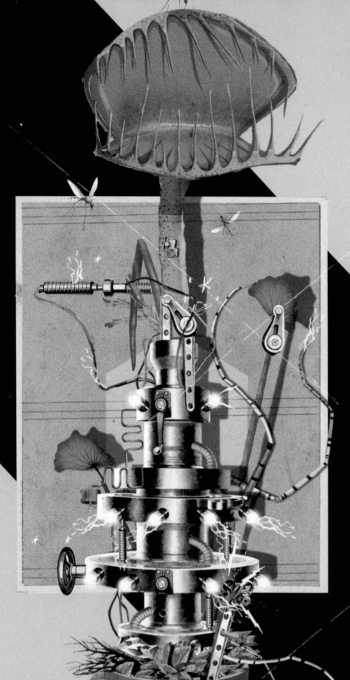

Matthew Billington

IN THIS FUN, EASY PROJECT YOU'LL create a remote-controlled Venus flytrap by connecting an Arduino microcontroller to specific locations on a leaf with homemade electrodes and learn how biological action potentials can be triggered with small electrical signals. It's affordable, takes less than 90 minutes to put together (plus around 24 hours for the flytrap to rest), and makes a great classroom demonstration. Add a wi-fi module and second circuit to sense when the leaf closes, and you've got an internet-connected, fly-detecting cyborg plant!

HOW IT WORKS

Much of the biological world is driven by the passive movement of matter from high concentration to low concentration. In live organisms such as our Venus flytrap, cells utilize these concentration gradients to move *ions* — molecules that carry a charge. Ca^{2+} (calcium) and K^+ (potassium) ions are positively charged; this means they need to gain electrons (e^-) in order to be neutral. Cl^- (chloride) is negatively charged, meaning it needs to give up an electron to be neutral.

When an insect stimulates the trigger hairs on the surface of a Venus flytrap, the cells of the trigger hair cause a change in ionic concentration. The corresponding change in electronic charge (e^-) causes the cells in the leaf's midrib to rapidly swell with water, causing the trap to close.

So what happens if we stick a Venus flytrap in the middle of a circuit? From the plant's perspective, electronic charge is dumped into the cellular space and it "feels" like a change in ionic concentration. In this way, we can use electricity to simulate a fly landing on the trap and trigger a biological response!

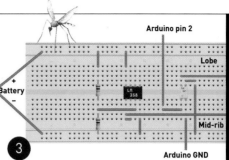

Arduino pin 2
Lobe
Battery
+
−
Mid-rib
Arduino GND

Gunther Kirsch

2

3

Time Required:
90 Minutes
Cost:
$30-$70

JORDAN HUSNEY
is strategy director for
Undercurrent. When he's not
consulting, he's connecting
everything and anything to
the internet. The BBQ grill is
still proving to be a challenge.
jordan.husney@gmail.com

DR. SASHA WRIGHT
is an ecologist at the Jena
Biodiversity Experiment,
where she designs computer
models to predict the effects
of climate change. She teaches
ecology and plant physiology
at Sarah Lawrence College.
sashajwright@gmail.com

1. MAKE THE ELECTRODES

Put the bleach in the shot glass and note its level. Twist each silver wire around the toothpick, leaving 1" or so to dangle, so that 1cm of one end may be submerged into the bleach.

Place the toothpick on top of the glass and allow the wires to react for about 15 minutes (**Figure 1**), forming a layer of silver chloride. They'll look tarnished when they're done.

Unwind the wires and rinse them well under a running tap.

2. PLACE THE ELECTRODES

Carefully insert one electrode through the midrib of the Venus flytrap leaf. It may close, which is OK. Insert the second electrode through the middle of one of the leaf's 2 lobes (**Figure 2**).

Separate and tape each electrode wire to the side of the plant's pot to reduce movement and prevent short-circuits.

3. BUILD THE CIRCUIT

Connect the AA battery holder's 2 wires to 2 separate buses on the breadboard. Then place the resistors, op-amp chip, optocoupler, and jumper wires as shown in **Figure 3**. Step-by-step breadboarding instructions and a schematic diagram are available on the project page at makezine.com/RCflytrap.

4. TEST THE CIRCUIT

Install 2 AA batteries in the battery holder and make sure it's properly wired to the breadboard.

Connect the USB cable between your computer and the Arduino. Download the Arduino sketch *VenusTrigger.ino* from the project page. Open the sketch in the Arduino IDE, and upload it to the board by pressing the arrow button. Open the Serial Monitor window, enter the text "hi" and press the Send button. You should see the text `Triggered!` followed by `Waiting for keypress:` in the window.

Touch the multimeter's red probe to the optocoupler's pin 1, and the black probe to pin 2. You should read 0V.

Send "hi" again using the Serial Monitor. The multimeter should read 5V for 5 seconds.

Next, touch the red probe to both the op-amp's free jumper wire and the optocoupler's free jumper wire. It should read 0V. Press another key in the Serial Monitor. It should read approximately 1.5V for 5 seconds. If so, you're ready to connect to a real plant!

5. ACTUATE YOUR ROBOTIC PLANT

Wait for the Venus flytrap leaf to fully open. (This may take between 12 and 72 hours — be patient.) Now it's time to impress your friends.

Connect the op-amp's free jumper wire to the midrib electrode. Connect the optocoupler's free jumper wire to the leaf lobe electrode. The leaf should remain open.

Back in the Arduino's Serial Monitor window, prepare to press a key. On the count of 3, say, "Fire in the hole!" and press any key. The leaf should close! If it doesn't, try zapping it again. Sometimes it takes 2 or 3 pulses. You now have robotic command of a flesh-eating plant!

EXPLORING POSSIBILITIES

Now that you've made your remote-controlled Venus flytrap, you've probably got a *million* ideas on how you're going to use it. But here are a couple suggestions:

» Consider using a wi-fi or XBee Arduino shield to make a **wireless Venus flytrap**.
» Use the electrodes attached to the Venus flytrap as a **cybernetic fly detector**. When the trigger hairs are stimulated, a very small voltage spike (only 0.14V) will briefly (<2ms) appear across the electrodes. A chip called an instrumentation amplifier (such as Analog Devices AD623) can take this tiny signal and amplify it hundreds or thousands of times so it can be detected by the Arduino. ⏻

Get the project code, breadboarding instructions, and video of the robo-flytrap in action at makezine.com/RCflytrap
Share it: *#makeprojects*

Materials

» **Venus flytrap plant** *Dionea muscipulosa* (var. "King Henry"), available from petflytrap.com
» **Silver wire, 99.99% pure, 5" length** (2)
» **Toothpick**
» **Shot glass**
» **Household bleach, 10mL**
» **Tape** scotch, masking, or electrical
» **Arduino Uno microcontroller** Maker Shed item #MKSP99, makershed.com
» **USB cable, Standard A to Standard B**
» **Solderless breadboard** Maker Shed #MKEL3 or #MKKN3
» **Jumper wires (15)** Maker Shed #MKSEEED3
» **Battery holder, 2×AA**
» **Batteries, AA** (2)
» **Resistors, 10kΩ** (2)
» **Optocoupler integrated circuit (IC), 4-pin DIP, 125mA** such as Lite-On LTV-814H
» **Operational amplifier IC, LM358**

Tools

» **Multimeter**
» **Computer with Arduino IDE software** free download from arduino.cc/en/Main/Software

SunBEAM SEEKER BOT

Written, photographed, and illustrated by Sean Michael Ragan

Our easy-bake adaptation of a classic phototrope.

THIS LIGHT-RESPONSIVE BEAM-STYLE ROBOT IS BASED ON RANDY SARGENT'S CLASSIC "HERBIE" DESIGN. Using as few as six components, the "Herbie" circuit produces a robot that will follow a black line traced on a white floor or, with the photosensors pointing upward as shown, seek out the brightest light source in a room. To the basic design, the SunBEAM Seeker adds a roller lever limit switch that serves as a tail-wheel so the bot will automatically turn on when you set it down, and off when you pick it up (or if it flips over).

1

2a

2b

3a

3b

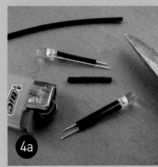
4a

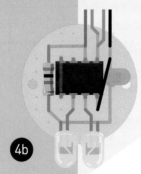
4b

5

1. THE GUTS

Bend the threaded "ears" on the slide switch back and forth with pliers until the metal fatigues and breaks. File off any sharp edges.

Clean the back of one battery pack and the sides of both switches with alcohol. Cut and apply small sections of double-sided foam tape to attach the switches to the battery pack. Center the roller switch along the bottom edge, and put the slide switch in the upper right corner about ⅛" back from the right end. Cut, strip, and solder the red battery pack lead to connect the 2 switches in series.

Cut the red lead off the other battery pack right where it connects to the terminal inside the battery holder.

2. THE BODY

Apply foam tape to the exposed sides of the 2 switches. Route the wires (**Figure 2a**) — one red lead goes out the side, and one goes out the top. Thread the black lead into the second battery holder through the opening where the red lead used to come out. Remove the foam tape's protective film and attach the battery holders back-to-back, with the switches sandwiched between them.

Cut, strip, and solder the black lead to the battery clip terminal *inside* the battery holder (where you removed the red lead earlier).

Clean the front side of the body with alcohol, then cut and apply a piece of Superlock fastener tape to fit it (**Figure 2b**).

3. THE LEGS

Clean the motor's flat side (without the vents) with alcohol, then apply 2 small strips of Superlock fastener tape. Use a hobby knife to cut 2 fresh pencil erasers off at the ferrules. "Stab" one eraser onto each motor shaft to serve as a wheel. Center the shaft in the eraser carefully or the wheel will wobble. Attach the motors to the front of the body using Superlock fasteners (**Figure 3a**).

Attach the front red lead to the outboard terminal on the "starboard" motor. (We covered our red lead with black heat-shrink tubing, just for aesthetics.)

Attach the black lead to the inboard terminal on the "port" motor. Attach a second black lead, made

CAUTION: Overheating the terminal while soldering may melt it loose from the plastic case.

TIP: Once the foam tape is in place, you may want to color the edges with a marker just for looks.

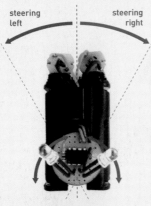

steering left · steering right

faster · faster

slower

from about 4" of leftover battery holder wire, to the same terminal. Leave its other end loose for now.

Attach one lead (4" or longer) to each of the 2 remaining motor terminals. These will both connect to pin 5 (**Figure 3b**).

4. THE HEAD

Cover the leads of both IR detectors with heat-shrink tubing except for about 5/16" (on the longer lead) at the ends (**Figure 4a**). The leads function as maneuverable "eyestalks," and insulating them above the PCB keeps them from shorting if they get bent or twisted together.

Solder the chip, the resistor, the LED, the black jumper wire, and the two IR detectors to the PCB as shown in **Figure 4b**.

5. IT'S ALIVE!

Thread the loose wires from the motors into the body in front of the slide switch, then back out the top right behind it. Cut, strip, and solder the wires — power (red), black (ground), and the 2 "shared" motor leads (green) — to connect the PCB.

Apply double-stick foam tape "ears" on the underside of the PCB — one to either side of the solder traces. Cut the tape to match the profile of the PCB. Clean the top of the body with alcohol, then peel the tape backing and press the PCB into place.

Make sure the slide switch is in the rearward "off" position, then install 4 AAA batteries in the holders. Flip the switch forward and set your bot down on a smooth floor. If it's working right, the robot will come alive and speed off happily in search of the brightest light in the room.

CARE AND FEEDING

The angle of the motors affects drive speed and power. A steeper angle is slower but stronger, and a shallower one faster but weaker. If your bot prefers turning one way over the other, the "eyes" can be adjusted left and/or right to bias steering and compensate as needed. ◐

For video, schematics, and other resources, please visit makezine.com/sunbeambot
Share it: **#makeprojects**

Time Required:
2 Hours
Cost:
$20-$30

Materials

» **Battery holder, 2 × AAA (2)** RadioShack #270-398
» **Slide switch, SPST subminiature** RadioShack #275-032
» **Roller lever switch, SPDT with ¾" lever** RadioShack #275-017
» **Double-sided tape, foam** RadioShack #64-2343
» **Superlock fastener strip** RadioShack #64-2363
» **DC motors, 6VDC (2)** RadioShack #273-106
» **Pencil erasers (2)**
» **Hookup wire, 24—22 AWG, 6"** I used one of the long green breadboard jumper wires from RadioShack #276-173.
» **IR detectors, phototransistor type (2)** RadioShack #276-142
» **Heat-shrink tubing, 1/16" nominal** RadioShack #278-1627
» **LM386 low-voltage power amplifier chip, 8-pin DIP** RadioShack #276-1731
» **1K Resistor, ¼W** RadioShack #271-1321
» **LED, 3mm** RadioShack #276-026 (red) or #276-069 (green)
» **1" Round PCB** RadioShack #276-004
» **Batteries, AAA (4)** RadioShack #23-850

Tools

» **Pliers, mini long-nose**
» **File**
» **Scissors**
» **Rubbing alcohol**
» **Wire stripper/cutter**
» **Permanent marker (optional)**
» **Soldering iron**
» **Solder**
» **Heat gun, hair dryer, or butane lighter** for heat-shrink tubing
» **Hobby knife**

MINIBALL SOLAR HAMSTERBOT

Hours of fun *sans* batteries.

Written by
Jérôme Demers

WHEN I STARTED IN BEAM ROBOTICS IN THE LATE '90S, the most popular robots were the rolling ones — solarrollers, photovores, symets, and the original miniball. The miniball was relatively unpopular because, unlike other BEAM bots, its "guts" were hidden under the solarcell and you couldn't see how it worked. It moved by shifting its internal center of mass, which was mechanically complex to pull off.

The idea of building a miniball roller inside a transparent sphere — like a hamster in a ball — and replicating the motion without using a complex gearbox, occurred to me in 2009 while I was working as an intern at Solarbotics. Instead of using a gearbox and a shifting weight, the solar miniball uses a simple direct-drive friction-wheel arrangement, with the inside diameter of the ball serving as a kind of "spherical gear" to reduce speed and increase torque. ◐

[For step-by-step instructions and more BEAM robotics projects, check out makezine.com/hamsterbot
Share it: *#makeprojects*]

Damien Scogin

THE MILLER SOLAR ENGINE

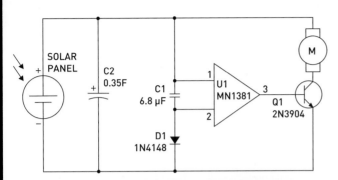

This robot uses a solar engine circuit. A small solar cell cannot provide enough power to constantly run a motor, so this circuit stores power in a capacitor (C2), then releases it in bursts to the motor.

Time Required:
2 Hours
Cost:
$20–$30

JÉRÔME DEMERS works in R&D for a leading company in the powersports industry. He is passionate about robotics and embedded systems. At Robowars 2014, he took first place in the 3kg autonomous sumo event in Montréal. See more of his robotics and electronics projects at jeromedemers.com.

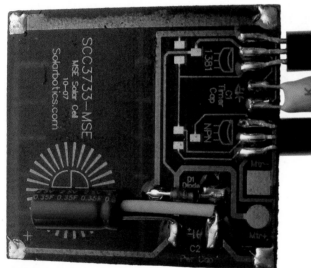

Jérôme Demers

Originally designed by Andrew Miller in 1995, the rights to this circuit were acquired by Solarbotics. The SCC3733 solar panel comes with ready-made printed circuit art, on the back, for assembling the Miller solar engine. If you are using another panel, you can "free-form" the solar engine. See makezine.com/hamsterbot for more info.

Materials
» **80mm clear plastic sphere** from craft or hobby shop
» **6mm×12mm 3V DC pager motor with leads** with eccentric weight removed
» **Fuse clip to fit motor**, such as Littelfuse #102071
» **Rubber wheels with nylon hubs** such as Solarbotics #RW
» **Paper clips (2)**
» **Small solar cell rated 5V at 20mA or better** such as Solarbotics #SCC3733
» **0.35F 2.5V double-layer "gold" supercapacitor**
» **6.8μF tantalum or electrolytic capacitor**
» **NPN transistor** 2N3904 or equivalent
» **MN1381 voltage trigger** or equivalent
» **Signal diode 1N914/1N4148**
» **Scrap breadboard jumper**

Tools
» **Soldering equipment**
» **Needle-nose pliers**
» **Flush cutters**
» **Safety glasses** are not optional when clipping wire.

BIOLOGY, ELECTRONICS, AESTHETICS, MECHANICS

BEAM is a way of thinking about and building robots with roots in the "behaviorist" or "actionist" robotics movement of the 1980s. Rather than relying on microprocessors, programming, and digital logic, BEAM designs favor discrete components, stimulus-response control systems, and analog logic. From a design perspective, BEAM robotics is about getting the most complex and interesting behaviors using the simplest circuits, actuators, and components.

Beetlebot photo: Jérôme Demers;
Pummer photo: Zach DeBord/M27.com

Beetlebot, Jérôme Demers, MAKE Vol. 12

Solarroller, Gareth Branwyn, MAKE Vol. 06

Pummer, Zach DeBord, MAKE Vol. 08

WORLD'S SMALLEST LINE-FOLLOWING ROBOT

Written by
Naghi Sotoudeh

Fabricate an incredibly small and capable vibrobot using a handful of components.

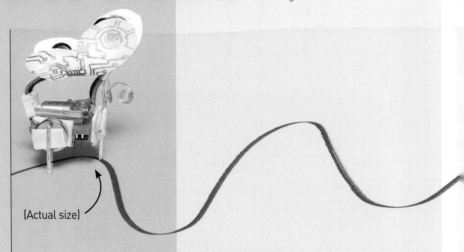

(Actual size)

RIZEH IS A PERSIAN WORD THAT MEANS "TINY," AND OUR RIZEH ROBOT IS VERY SMALL INDEED. Driven by two cellphone vibrators, it's very inexpensive to build and operate. The robot is able to execute basic linear and circular motions, and with its downward-facing infrared detectors, can accurately follow a black line on a white background. To our knowledge, it's the smallest line-following vibrobot ever made!

We applied some clever design tricks to minimize the number of components needed to control the robot, which is crucial when you're building at this scale. Instead of a motor driver, our control scheme exploits the built-in PWM capabilities of the microcontroller itself to control the vibrators. The complete robot weighs just 5 grams and measures 19mm × 16mm × 10mm — that's less than one quarter of a cubic inch.

We've tested and proven this design on line-following courses that include curves, lines, and corners. In 2013, Rizeh won first place in the Demo (Freestyle) League at RoboCup IranOpen.

Here's how to make your own.

Special thanks to Prof. Adel Akbarimajd, my partner in this project.

3a

3b

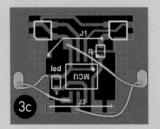

3c

5

1. MAKE THE PCB

Download the printed circuit board (PCB) files from the project page at makezine.com/robo-rizeh and prepare your PCB. We've provided DXF files for milling and Gerber files for etching (see makezine.com/primer-printed-circuit-board for a tutorial on PCB etching). This one was milled on a ShopBot Desktop at MAKE Labs.

2. PROGRAM THE MICROCONTROLLER

Download the code from the project page, then upload the .hex file to the ATtiny microcontroller using AVRDUDE or similar software, plus a hardware AVR programmer like Adafruit's USBTinyISP. (If you haven't done this before, you can follow Adafruit's tutorial at learn.adafruit.com/usbtinyisp/avrdude.) You'll also need to connect the ATtiny to an 8SOIC-to-8DIP breakout board for programming; those pins are tiny!

3. ASSEMBLE THE CIRCUIT BOARD

Bend or cut the microcontroller pin 1 (reset) so it won't contact the ground trace on the PCB (**Figure 3a**). Then assemble and solder the components in place on the PCB: the microcontroller, IR sensors, resistor, LED, and pin header. Use double-stick tape to mount the battery (**Figure 3b**). You should be able to plug its 2mm JST connector into your pin header with a little wiggling.

Solder the leads for each vibrator motor as shown (**Figure 3c**), paying close attention to wire

Time Required:
A Weekend
Cost:
$25–$40

NAGHI SOTOUDEH
is a robotics, automation, and electronics designer from Ardabil, Iran. He earned a BS in electrical engineering from Islamic Azad University in 2009. He now lives in Tehran and works as a circuit designer and programmer for Faratel Electronics.

Materials

» **Atmel AVR ATtiny45 microcontroller chip**
» **Infrared (IR) emitter/ detectors, compact DIP type (2)** such as Fairchild QRE1113 or Sharp GP2S60
» **LED, surface mount (SMD) size 0805**
» **Resistor, 100Ω, SMD size 0805**
» **Cellphone vibrator motors, 3V, coin shape, 10mm×2mm (2)** Parallax #28821
» **Battery, 3.7V, LiPo, 12mm × 16mm, with charger** SparkFun #PRT-11316 and PRT-10217
» **Pin header, 2mm pitch, male, 1×2** You can cut this down from larger 2mm header blocks.
» **Wire, solid insulated, fine**
» **Foam tape, double-sided**
» **Pins, needles, or brads (3)**
» **Craft foam, sheet**
» **Black marker, 6mm tip**

Tools

» **Computer with AVR programming software** such as AVRDUDE
» **AVR programmer, USB** such as Adafruit USBtinyISP
» **Breadboard**
» **Jumper wires**
» **Breakout board, 8SOIC to 8DIP** Digi-Key #309-1098-ND
» **Soldering iron and solder**
» **Wire strippers / cutters**
» **PCB milling or etching gear**

---

colors and polarity. For the robot's right motor (shown left in the diagram), solder the blue negative (–) wire to the ground trace right beside the LED, and the red positive (+) wire to the trace leading to microcontroller pin 5. For the robot's left motor (shown right), solder the blue negative wire to a trace leading to microcontroller pin 6 and its red positive wire to the large ground trace above microcontroller pin 4 (ground). Finally, solder jumpers wires J1 and J2 as indicated.

4. MOUNT THE VIBRATORS

Remove the adhesive tape backing from the vibrator motors and mount them to opposite edges of the circuit board as shown. You can use additional foam tape for strength (or wrap them instead with 2 turns of single-sided tape), but take care not to block the sensors.

5. ADD THE LEGS

Use your wire cutters to trim 2 pins, needles, or brads down to 12mm, and 1 more to 13mm. With the PCB positioned component-side up, tape a 12mm leg to each vibrator, centered with its point straight up. Mount the 13mm leg, centered to the robot's "head" using more tape or a dab of epoxy. Your build is now complete. (You can add ornamental butterfly wings if desired.)

6. BUILD YOUR TEST TRACK

The track surface needs to be made from soft material so the legs get traction and don't skip around too much. Craft foam works very well. Draw the route you want the robot to follow on a white sheet of it using a black marking pen with a 6mm-wide tip.

7. TRY IT OUT!

Plug in the battery to turn your robot on. Hold it upside down in the air for a few seconds to calibrate the sensors, then set it down astride the line you drew and watch it go! If it's skittish, try adjusting the length of the "head" pin or shifting the weight forward by moving the battery.

HOW IT WORKS

The microcontroller compares the 2 sensor inputs to see which one is detecting a brighter background, and activates the motor on that side. The triangular arrangement of the pins causes the vibrating side to "walk" forward, rotating the robot around the head pin toward the opposite side. Result: The robot continually moves forward, always redirecting itself toward the center of the dark line. For more information and video, visit roborizeh.ir and read our paper in Advanced Robotics, "Design and motion analysis of vibration-driven small robot Rizeh." ◐

Get the PCB files, project code, and more build photos at makezine.com/robo-rizeh
Share it: *#makeprojects*

MARS-BOT:
ADDING SCIENCE TO ROBOTICS

A simulated space mission could leverage the popularity of sport robotics to teach science.

Written by Forrest M. Mims III

SCIENCE FAIRS THAT INVOLVE ENGINEERING, PHYSICS, ASTRONOMY, AND CHEMISTRY HAVE DECLINED IN RECENT DECADES, while robotics competitions have rapidly grown in popularity. These contests teach students about electronics, mechanical engineering, and teamwork while providing plenty of fun in the process.

After watching a number of robotics competitions, I'm confident they can be expanded to include some science. So here's my proposal for a new kind of robot competition: Mars-Bot, a simulated space mission.

DESIGNING A MARS-BOT COMPETITION

A suite of sensors and samplers will transform a standard competition robot into a sophisticated Mars-Bot that is guided across an imitation Martian landscape by a student mission team (**Figure A**). Each Mars-Bot will be required to perform a minimum number of tasks after it disappears behind a dark curtain concealing the Martian landscape from the mission team. Only the audience on the other side of the curtain will be able to see the Mars-Bots — each team must rely on video images sent back by its Mars-Bot to navigate across the Martian terrain, execute its mission, and return through the curtain (**Figure B**).

Before the competition, each team receives an image of the Martian landscape, as if taken from a satellite, that identifies landmarks. A grid laid over the image provides dimensions, and the bot must visit and sample or measure the landmarks. If the competition is indoors, a bright overhead light can represent the sun. Scoring is based on the number of successful measurements and samples each Mars-Bot returns, with ties broken by speed. Each team will create comprehensive mission reports and data analysis to receive academic credit.

VIDEO CAMERA(S)

The most indispensable sensor on each Mars-Bot will be at least one wireless, color webcam. The basic setup has a stationary mount that looks straight ahead. More sophisticated

Jim Burke

A Mars-Bot will include the drive and steering mechanisms of a conventional competition robot, plus a wireless video camera and various sensors.

FORREST M. MIMS III (forrestmims.org), an amateur scientist and Rolex Award winner, was named by *Discover* magazine as one of the "50 Best Brains in Science." His books have sold more than 7 million copies.

Mars-Bots will feature a webcam that can rotate to better survey the landscape, find assigned goals, and provide visual clues when samples are collected. The camera itself can also report back data indicated by readouts or instruments in its field of view.

SIMULATED DUST STORM

Mars is known for its vast dust storms. A fan blowing dust across the path of a Mars-Bot could test the ability of moving parts to survive a blast of grit. The reduction in electrical power that occurs when dust falls on a Mars-Bot's solar panel can also be measured.

WIND-SPEED SENSORS

A Mars-Bot should measure the speed of any wind it encounters. A fan or propeller, mounted on the shaft of a small DC motor, can act as a simple analog wind speed sensor. When wind rotates the motor's armature, a voltage proportional to the rotation rate will appear across the motor's terminals.

A digital wind speed sensor can be made by mounting a disk on the shaft of a propeller. Glue a small magnet to the outer edge of the disk, and mount the assembly so the edge of the disk rotates past a Hall effect sensor. The Hall sensor will provide a voltage pulse each time the magnet passes by. If needed, additional magnets or weights can be mounted around the disk to balance it. Calibrate the sensor by placing it adjacent to a commercial, handheld wind-speed sensor at various distances from a fan.

HAZE

Dust blown high into the Martian atmosphere can cause long-lasting haze. In an indoor competition, periodic dimming of the artificial sun can simulate haze, while passing clouds will create the same effect during an outdoor competition. A photodiode or solar cell can detect the reduced light. Mount it behind a plastic diffuser to ensure it receives light no matter the artificial sun's location.

TEMPERATURE

The temperature during a mission will slightly change with wind, haze, and cloud conditions. It can be easily measured using a thermistor or integrated temperature sensor, or better yet, with an infrared thermometer, which can also scan the temperature of various objects along the mission course.

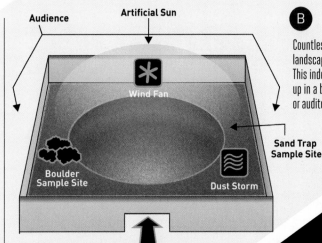

Audience — Artificial Sun

B

Wind Fan

Boulder Sample Site

Dust Storm

Sand Trap Sample Site

Curtained-Off Mars-Bot Preparation, Launch, Recovery and Analysis Area

Countless simulated Mars landscapes can be imagined. This indoor version can be set up in a basketball court or auditorium.

SPECTROMETER

The colors of rocks, sand, and soil provide important clues about their composition. The Mars-Bot's video camera can be used as a simple 3-color spectrometer. Photo processing software can analyze individual video frames to express the relative intensity of the blue, green, and red wavelengths of the simulated Martian landscape.

SAND AND PEBBLE SAMPLING

An especially important part of a Mars-Bot mission is to collect geological samples and return them for analysis. Mission controllers will use their video link to steer their Mars-Bot to sand and gravel sites along the course. The mechanical features of the current generation of competition robots can be easily modified for sand and pebble sampling.

BORER TO COLLECT "ROCK" SAMPLE

The mission team will steer their Mars-Bot to a mock boulder, bore a sample, and stash it for the return trip. The boulder might be fashioned from a thin but rigid sheet of wood or other soft material. A standard battery-powered drill fitted with a ½" to 1" hole saw can be mounted on the front of the Mars-Bot and switched on and off by a radio-controlled relay connected across the drill's power switch. In my experience, a circle of wood removed by a hole saw stays inside the saw until it's manually removed, so the saw itself should hold and retain one or two thin samples.

These are just a few ideas; I would love to hear yours. I'm hopeful that science-oriented Mars-Bots will significantly expand the hands-on educational experience already provided so well by established robotics competitions. ⊘

Forrest M. Mims III

Robotics competitions show how students can learn about electronics, mechanics, and teamwork while having fun in the process.

Post your Mars-Bot competition ideas at makezine.com/marsbotics Share it: *#marsbotics*

OPEN-SOURCE ROBOTICS

A buyer's guide
Written by Sean Michael Ragan

THE PHRASE "OPEN SOURCE" DESCRIBES A PRODUCT WHOSE DESIGN is published with the understanding that anyone is legally free to modify, distribute, make, or even sell their own version of it. In the case of hardware like robots, the same design will often be available for sale — perhaps with various modifications and in various stages of completion, from parts to kits to fully-built — from unaffiliated manufacturers. Some of these may seem pricey, but remember you're free to make, mod, or even manufacture as much (or as little) of the finished product as you want.

DARWIN-OP

$12,000 trossenrobotics.com

The Dynamic Anthropomorphic Robot with Intelligence — Open Platform is a cutting-edge research humanoid created by Virginia Tech's Robotics and Mechanisms Laboratory in collaboration with the University of Pennsylvania, Purdue, and Robotis. Hobbyist Michael Overstreet has built his own working DARwIn-OP, ("How I Printed a Humanoid," MAKE Volume 34, page 66), using the open-source files, for about half the sticker price.

[For more cool open source robotics projects, commercial and otherwise, please check out makezine.com/oshwbots Share it: *#oshwbots*]

BUDDY 4000

$45 (frame) robotgrrl.com

One of several low-cost platforms designed by Erin "RobotGrrl" Kennedy, Buddy 4000 features a stylish 3D-printed body with an Arduino Pro Mini for a brain, independent arm and neck servos, and 6 LED headers. Download and print the frame for free, or purchase a preprinted kit from RobotGrrl herself.

UARM

$300 ufactory.cc

Billed as "a miniature industrial robot arm on your desk," the uArm is based on ABB's IRB 460 high-speed "palletizer." The design concentrates mass in the base, allowing the arm to move quickly. Spectacularly overfunded on Kickstarter in March, uArm can be preordered now and is expected to ship in June.

3DR PIXHAWK

$280 3drobotics.com

The successor to 3DRobotics' flagship ArduPilot Mega flight microcontroller, the Pixhawk offers all the capabilities of the original APM, plus an advanced ARM Cortex M4 processor, built-in MicroSD drive, and much more. Pixhawk ships with all 3DR's "one-box" multicopter drones including the IRIS, the Y6, and the X8 (shown here).

OPENROV 2.6

$850 openrov.com

The latest version of Eric Stackpole and David Lang's OpenROV project features water-clear, water-tight acrylic frame construction, a custom microcontroller and interface board, an HD webcam, high-intensity LED lights, dual laser emitters, and more. Power is supplied by on-board rechargeable batteries, and control via a lightweight 2-wire tether cable.

TURTLEBOT 2

$1,400 iheartengineering.com

TurtleBot cleverly combines several off-the-shelf systems to create a powerful, relatively low-cost platform with minimal assembly. There's a Kobuki wheeled base for mobility, a Kinect sensor for 3D vision, and a netbook running ROS (Robot Operating System) for processing and control. The TurtleBot software stack enjoys a very active volunteer developer community.

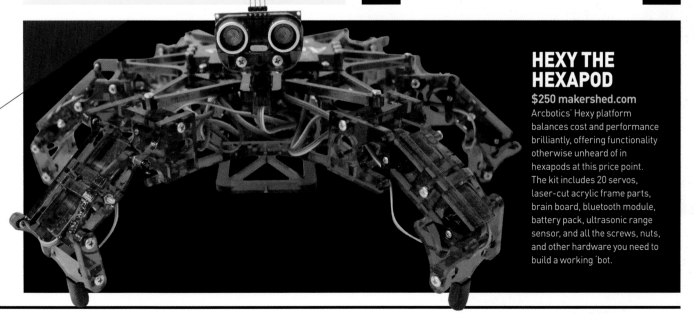

HEXY THE HEXAPOD

$250 makershed.com

Arcbotics' Hexy platform balances cost and performance brilliantly, offering functionality otherwise unheard of in hexapods at this price point. The kit includes 20 servos, laser-cut acrylic frame parts, brain board, bluetooth module, battery pack, ultrasonic range sensor, and all the screws, nuts, and other hardware you need to build a working 'bot.

SKILL BUILDER+

Learning new tricks every issue
Tell us what you want to learn about: *editor@makezine.com*

EASY

Photo Etching METAL PARTS

Use chemical milling to cut intricate shapes from brass or stainless steel.

Written, illustrated, and photographed by Bob Knetzger

WHILE WORKING ON THE "DESKTOP FOUNDRY" project for MAKE Volume 36 (makezine.com/desktop-foundry), I was surprised to learn that my local shops with CO_2 laser cutters wouldn't cut brass. It's too reflective and can damage the lenses. What to do? Instead, I thought I'd try to create the intricately detailed brass parts using a different process: photo etching. Better living through chemistry!

Photo etching is used to make parts for small-scale model trains. This 1:48 scale model helicopter from Czech Republic comes with a sheet of photo-etched brass parts **(Figure A)**. Czech out the exquisite detail! Those tiny bolt heads are only 0.010" in diameter.

British designer Sam Buxton has created spectacularly clever art pieces for his line of Mikro Men **(Figure B)**. A single flat piece of etched stainless steel is bent and twisted to create a miniature three-dimensional world. My favorite is a bubble-helmeted space man taking his pet Mars Rover out for a walk, holding a robotic pooper-scooper. There are also half-etched details of textures, lettering, and markings.

I photo etched parts for the Desktop Foundry project **(Figure C)**. Some of the pieces were flat, decorative brass trim but two parts were bent and folded to create

Time Required:
2–3 Hours
Cost:
$140–$160

Materials

» **Metal photo-etching kit** Micro-Mark Pro-Etch #83123, micromark.com, includes: clear inkjet printer sheets, plexiglass squares, roll of photoresist film, laminator, carrier sheets, ferric chloride solution (40wt%), etching tank, spring clamps, tubing clamp, rubber airline, air pump, sodium hydroxide solution (5wt%), polishing pad, brush, 0.005" brass and stainless steel sheets, trays, aerator nozzle, gloves, goggles, measuring cups, and an excellent instruction manual. (See review on page 61)

A

B

C

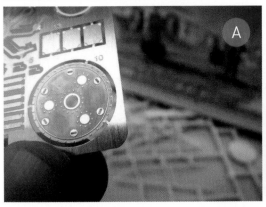

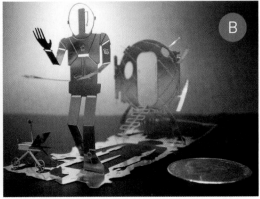

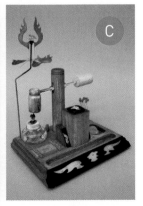

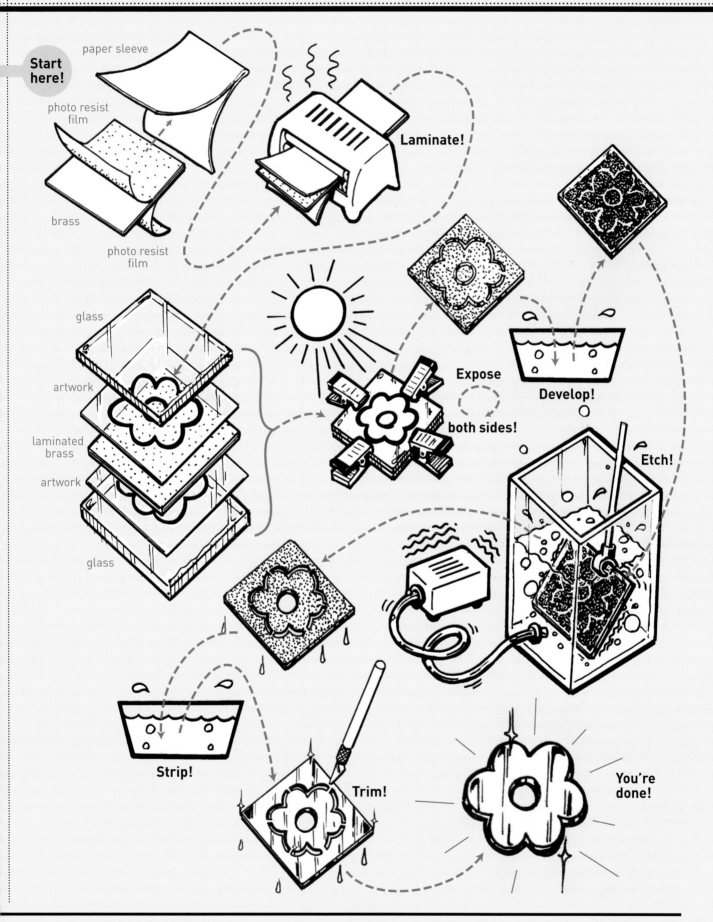

Start here!

paper sleeve

photo resist film

brass

photo resist film

Laminate!

glass

artwork

laminated brass

artwork

glass

Expose both sides!

Develop!

Etch!

Strip!

Trim!

You're done!

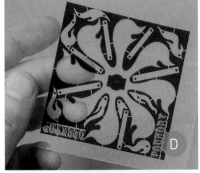

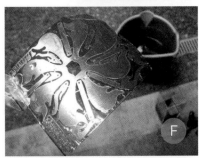

a three-dimensional phoenix-shaped turbine.

If you've ever etched your own circuit boards you may already have most of the required materials. If not, consider using the excellent Pro-Etch kit from Micro-Mark. It has everything you need, including inkjet printer sheets, photoresist film, metal sheets for etching, etchant chemicals, and special tools.

The process is similar to etching a printed circuit board, except there's no board. In short: Print your image on clear film and place it over the photosensitized brass. Expose it to a light source and place it in developer. The areas covered by the black image wash away, leaving an acid-resistant pattern. Place the brass in ferric chloride solution to etch away the unprotected areas, leaving your etched part. Here are the details.

Create Your Image
Photo etching can create very small details and features, down to a tiny 0.010" or so.

However, the smaller the feature, the more care you'll have to take in keeping the image razor sharp and the exposure blur-free. Also, the finer the detail, the more vigilant you'll need to be when developing and etching.

Create your image as a bitmap in Photoshop or vector art in Illustrator, or even just go old school with hand-drawn images. Add dotted lines to aid in bending or folding the finished shape. Connect all of the parts and shapes with thin lines so, when etched, they stay together and don't fall off into the tank of etchant. Leave a generous border of non-etched brass around the parts to act as a frame to hold it all together while etching. Print out your image on clear film with dark black ink **(Figure D)**. The clear parts of the image will not etch — those are the resultant metal parts.

Unlike etching a PC board, in this process the acid will etch thru both sides of the brass at the same time. You can choose to have a solid area of resist to protect the back, result-

TIP:
KEEP IT SIMPLE. ON THE FAN BLADE ARTWORK I EXPERIMENTED WITH SOME FINELY DETAILED PARTS AND FILIGREED LETTERING. BAD IDEA — MUCH OF THE DETAIL "BLEW OUT" DURING DEVELOPING AND ETCHING. GO FOR COOL DETAIL BUT DON'T OVERCOMPLICATE YOUR PARTS.

ing in a simple part. Or you can have slightly different images on the front and back.

Where the areas of image are the same, the brass is etched completely through, making a hole. Where the images are different, etching only happens halfway into the metal on one side, making an engraved feature like the tiny bolt heads or markings in the previous examples.

Transfer to Metal
First prepare the surface of the metal. It must be grease-free and perfectly clean. Scrub with a moistened polishing pad and handle the metal by the edges only — no fingerprints! Peel off the protective carrier and adhere the photo-resist film to the metal with a little water. Smooth out any trapped air bubbles and excess water. Repeat for the other side. Then place the film/brass/film sandwich in a protective paper cover sleeve and run it through the laminator. The heat and pressure adheres the photo film to the brass.

Place the printed image over the brass, clamp between plexiglass, and expose to UV light. Use the bright noon sun for 15 seconds or a 100-watt bulb with a longer exposure time. When properly exposed, you should see a dark purple image — the negative "shadow" of your artwork. Repeat the exposure for the back, which for my part had no image (i.e. solid resist).

Remove the top layer of clear protective films from both sides and immerse the brass in a diluted developer bath of 10 parts water to 1 part sodium hydroxide. The unexposed areas will soften and wash away, exposing the bare brass underneath. Gently brush the surface to help remove the softened film until you have crisp edges on all the smallest features **(Figure E)**. Rinse clean with water.

TIP:
BE EFFICIENT. MY FAN DESIGN HAD MUCH MORE EXPOSED AREA TO BE ETCHED AWAY THAN IT REALLY NEEDED. THAT DEPLETES THE ETCHANT SOLUTION FASTER AND SLOWS DOWN THE PROCESS. YOU REALLY ONLY NEED TO ETCH AWAY A THIN OUTLINE AROUND YOUR SHAPES — NOT THE ENTIRE SHEET.

Etch

Immerse the brass completely in a tank of ferric chloride. Hang the brass from a clip and periodically lift the part out of the tank to check its progress. The etchant is nasty stuff so wear protective gloves and goggles, and work in a space with good ventilation. For best results warm up the etchant solution in a bath of hot water and use an aerator to provide a flow of fresh etchant and gentle bubbling agitation.

Carefully brush away spent etchant from small details to speed up the process. Depending on the amount of metal to be etched and the level of detail, it may take 20 minutes or more to com-

> **TIP:**
> KEEP IT MOVING. CHECK ON YOUR PARTS IN THE ETCHING TANK EVERY COUPLE OF MINUTES. ROTATE THE ORIENTATION OF THE PART OFTEN, AND CHANGE ITS POSITION IN THE TANK TO ENSURE THE ENTIRE SURFACE IS GETTING A GOOD FLOW OF FRESH ETCHANT.

pletely etch your design **(Figure F)**.

Rinse the finished part under warm water to remove all the etchant. Return it to the tank if it needs more time. When done, dip the finished part in a tray of undiluted 5wt% sodium hydroxide to dissolve the purple resist material. Gently brush away any stubborn resist, then rinse in clean water **(Figure G)**.

Trim the parts from the supporting brass runners using tiny snips or an X-Acto **(Figure H)**. Carefully burnish with a polishing pad for a shiny look. For my phoenix-shaped part, I tapped the center with a center punch to make a slight dimple to act as a bearing. Then I folded it to make the wings and head of the phoenix, and twisted the blades to make the turbine fan. When assembled, the phoenix/fan rests on a sharpened rod and spins freely on the slightest bit of rising hot air. See it in action online at makezine.com/desktop-foundry.

I can't wait to try my next photo-etch project — it's fun! ◔

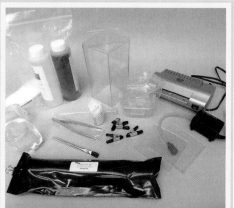

THE MICRO-MARK PRO-ETCH SYSTEM

$134 : micromark.com

This great kit has absolutely everything you'll need to photo etch at home, right down to gloves and safety goggles. The detailed instruction manual covers every aspect of the process clearly and is full of great tips and examples. I really like the custom laminator: The motorized heated rollers apply just the right heat and pressure, for just the right time to adhere the photo-resist film perfectly to the brass or steel. Also included is a tall etching tank with a clever clip support that suspends parts up to 3"×3" inside the tank. An aquarium pump with aerator "bubbling nozzle" agitates the sodium hydroxide for better, faster etching. Though it costs more than $100, it's worth it. I'll use it again and again.

BOB KNETZGER
is a designer/inventor/musician whose award-winning toys have been featured on *The Tonight Show, Nightline,* and *Good Morning America.*

For video, more photos, discussion, or to show us your own photo-etched parts, please visit makezine.com/photo-etching

Share it: *#makeprojects*

EASY | THE ECLECTIC Electret MICROPHONE

Written and photographed by Charles Platt

PCB TRACES
RADIATING "FINGERS" CONNECT CASE TO GROUND

PCB SUBSTRATE
WITH HOLES FOR TRANSISTOR LEADS

PLASTIC CASE
HOUSES INTERNAL AMPLIFER

AMPLIFYING TRANSISTOR
SILICON JFET N-CHANNEL TYPE, WITH GROUND ON SOURCE, PICKUP PLATE ON GATE, AND SIGNALOUT ON DRAIN

PICK-UP PLATE
FORMS CAPACITOR WITH MEMBRANE

PLASTIC SPACER
MAINTAINS INSULATING GAP

ELECTRET MEMBRANE
BONDED TO METAL

CASE
CRIMPED AT BACK TO RETAIN CONTENTS

DUST COVER
CLOTH OR PAPER

Time Required:
1 Hour
Cost:
$10–$20

Materials

» **Electret microphone** Most 8mm or 10mm electrets will work in this circuit, such as RadioShack #270-0090.
» **Ceramic capacitors: 0.1μF (2), 0.68μF, 10μF (2)**
» **Electrolytic capacitor, 330μF**
» **Resistors: 22Ω, 1K (2), 1.5K, 3.3K, 10K, 100K (2)**
» **Trimmer potentiometers: 10K, 100K**
» **Integrated circuits: LM741, LM386** Manufacturers may precede these generic part numbers with additional letters or numbers.
» **Loudspeaker: 2-inch or 3-inch, 5Ω–100Ω**
» **9V batteries (2) and connection clips**

THE MICROPHONES IN OLD-FASH-IONED WIRED TELEPHONES were relatively heavy, large, and expensive, and their sound quality was terrible. Thanks to materials science developments in the 1990s, electrets are now tiny, high-quality, and available from some sources for less than $1 each.

The electret microphone performs its magic with a pair of thin electrostatically charged membranes. When sound waves force one closer to the other, a tiny transistor in the microphone amplifies the fluctuations in electrical potential. We can amplify them further, in our circuits, and use them for many purposes.

Testing, Testing ...

Most electrets have two terminals. They may have leads attached or just solder pads for surface-mount applications. Since the pads are reasonably large, you can easily add your own leads if necessary.

Your first step is to distinguish the positive and negative terminals. They are not usually marked in any way, and the datasheets can be surprisingly uninformative. However, if you look

at the back of the microphone, you should see metal "fingers" radiating outward to the shell from one of the terminals (**Figure A**). These "fingers" — embedded in a translucent sealing compound — identify the negative side of the electret, which should be connected to ground.

Connect the other terminal through a 3.3K series resistor to the positive side of a 9VDC power supply, and you should see the electret responding to sound when you apply a meter (**Figure B**). Don't forget to set your meter to measure AC, not DC. A range of 1mV to 40mV is typical.

Alternate schematic symbols for an electret microphone.

Ⓐ Three electret microphones viewed from below. Upper right: from RadioShack. Upper left: from mouser.com. Bottom: from allelectronics.com. The ground terminal is on the right-hand side in each case.

Amplification

We can use an op-amp to turn millivolts from the electret into volts. **Figure C** shows a circuit using the LM741. While many simpler circuits exist, this one minimizes oscillations and distortion. The LM741 outputs to an LM386, a basic power amplifier chip that can drive a small loudspeaker.

Notice we use a "split power supply" consisting of +9V DC, −9V DC, and 0V (represented by a ground symbol). To set this up, you can use a pair of 9V batteries in series, as shown in **Figure D**. But why is it necessary?

Consider how sound waves are created. All around us is static air pressure, which can be imagined as an absence of sound. When you speak, you create waves that rise above the ambient level, separated by troughs that drop below it. An amplifier must reproduce these fluctuations accurately, and relatively positive and relatively negative voltages are the most obvious way.

In fig. C, a 0.68μF capacitor couples the microphone through a 1K resistor to the op-amp. The capacitor blocks DC voltage, to stop the op-amp from trying to amplify it. But the capacitor is transparent to the alternating audio signal, which we *do* want to amplify. The mic signal induces fluctuations in a neutral voltage provided by a voltage divider, and the op-amp amplifies the difference between these fluctuations and a second input, which has a stable reference voltage.

This reference is created with negative feedback from the op-amp output, adjusted with the 100K trimmer. Negative feedback keeps the op-amp under control, so that it creates an accurate copy of the input signal. To learn more signal processing with op-amps, look out for *Make: More Electronics*, the sequel to my book *Make: Electronics* scheduled for publication in May 2014.

Making It Work

The two-battery split supply has some limitations. The output won't be loud, and it may be scratchy, especially if your two 9V batteries deliver unequal voltage. Use the 100K trimmer to minimize the distortion and the 10K trimmer to maximize the volume. You'll get better results if you have a proper split power supply, or two 12V AC adapters connected through separate 9VDC voltage regulators.

A 50Ω-100Ω loudspeaker is preferred. I got really good results when I used alligator clips to connect the output from the circuit to the mini-jack plug on my computer speakers, but if you make a wiring error, you may damage your speakers.

You may be interested in other ideas, such as using sound to switch on a light or start a motor. For this purpose, instead of an audio amplifier such as the LM386, the output from the op-amp can trigger a solid-state relay. Add a 100μF capacitor between the op-amp output and ground to smooth the signal (so that the relay doesn't "chatter") and adjust the 100K trimmer until the sensitivity of the circuit is appropriate.

An op-amp can feed its output (through a coupling capacitor) to a microcontroller input pin. You'll have to discover the digital value that the analog-digital converter inside the microcontroller assigns to various levels of sound, but after that you can program different outputs for different sound levels.

Analog audio circuits can be trickier than digital circuits. Learning how to use an electret is a great introduction! ◗

Read and discuss this article at makezine.com/electret.
Share it: *#makeprojects*

CHARLES PLATT is the author of *Make: Electronics*, an introductory guide for all ages. He has completed a sequel, *Make: More Electronics*, and is also the author of Volume One of the *Encyclopedia of Electronic Components*. Volumes Two and Three are in preparation. makershed.com/platt

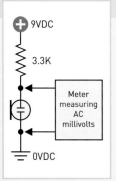

B Detecting the output from an electret microphone.

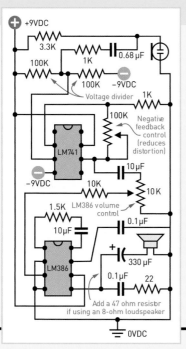

C An audio amplifier circuit.

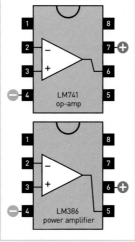

Basic pinouts of the LM741 op-amp and LM386 power amp. Unlabeled pins have additional functions; see datasheets for details.

D The split power supply required by the circuit can be provided by two 9V batteries wired in series.

To avoid oscillations and other noise, keep all wires as short as possible, and pack the components tightly together. The pairs of red and black wires connect with 9V batteries, while the yellow wires go to a loudspeaker.

Wood-Fired Barrel Oven

Build a super-efficient, easy-to-use backyard oven that'll never put cinders in your pizza.

Written by Max and Eva Edleson ▪ Photography by Eva Edleson
Illustrations by Max Edleson

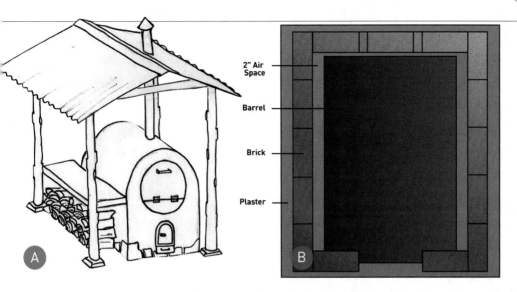

A

B

2" Air Space
Barrel
Brick
Plaster

**Time Required:
A Few Weekends
Cost:
$100–$1,200**

MAX EDLESON
is a professional artist/builder who is dedicated to using natural, local materials to create energy-efficient and spiritually uplifting elements of homes and public spaces, primarily masonry heaters and wood-fired ovens. He has a passion for farming, homesteading, and other traditional crafts.

EVA EDLESON
is a professional natural builder, cook, gardener, and craftswoman with more than a decade of experience specializing in natural wall systems, wood-fired cooking, and earthen paints and plasters. She has trained and worked with many of the most-respected natural builders in North America and Argentina.

A BARREL OVEN IS A VERSATILE AND HIGHLY EFFICIENT WOOD-FIRED OVEN THAT'S RELATIVELY EASY TO BUILD AND EASY TO USE. It can be the seed for a small-scale baking enterprise or the heart of a community's wood-fired cuisine. All kinds of food can be baked in the barrel oven, including bread, roasts, pizza, cookies, cakes, pies, casseroles, and stews.

The oven offers surprising convenience because it's hot and ready to bake within 15–20 minutes of lighting a fire, unlike traditional domed or vaulted pizza ovens that can take 2–3 hours! It's also easy to maintain at a desired temperature for long periods of time. With its highly conductive metal barrel surrounded by a thermal mass of masonry, this type of oven is often called a "mixed oven" because it has the capability to cook with direct as well as stored heat.

The barrel oven can be built from recycled materials or brand new parts. At its center is a steel barrel, with racks inside and a door at one end. Two deep shelves offer the ability to bake eight to ten 2lb (1kg) loaves of bread, four 12" (30cm) pizzas, or four cookie sheets at a time.

The secret to the barrel oven's efficiency is in its construction. The firebox is located beneath the barrel. The fire hits the bottom and wraps tightly around the barrel as it travels through the carefully constructed air space between the metal barrel and the surrounding bricks. This extended contact between fire and metal concentrates the heat for cooking inside the barrel and is what allows the oven to heat up so fast.

Since ash and carbon are not introduced to the cooking chamber, it stays clean and you can use baking pans interchangeably with other ov-

ens. The wood-fired "smoky" taste is not present in the food cooked in it.

Building a barrel oven is a manageable project for experienced builders or beginners. Once made, it becomes a center point for good meals and good times. Your barrel oven could easily last for many generations.

1. Choose a site

Plan the area around the oven as a gathering place, so you can both tend to the cooking and participate in the entertainment. The easier it is for you to engage with your oven, the more likely you'll use it. Design an outdoor kitchen site with:
» a rough footprint of 38" wide × 42" deep for a 55gal barrel oven
» countertops for food preparation and service
» easy access to your indoor kitchen
» storage for some firewood right next to the oven, and a pathway to your main wood storage.

2. Plan a roof

A roof is essential to protect the oven, keep it dry, and offer a place to cook in rainy weather. We recommend covering enough space for a small gathering of people.

You can create a small, independent roof structure that could also include work counters, a sink, benches, and wood storage (**Figure A**).

Or place the oven so the firebox and oven door are part of an indoor kitchen wall and the body of the oven is protected by a simple shed roof outside. This lets you use your oven indoors.

3. Lay out your oven

Draw on paper your barrel dimensions, and work out from there, including the width of the airspace and your bricks

Materials

There are many ways to build a barrel oven. Here's a "ballpark" materials list based on successful builds.

BARREL, DOORS & DRAWERS

» **Steel barrel, 55gal, "open head" type, clean** Industrial Container Services (iconserv.com) cleans and re-sells barrels. Or find your own from backyards, salvage yards, or food-industry locations.

» **Sheet metal, angle iron, welded wire grid, hinges, and handles** for fabricating the oven door, racks, ash grate, firebox, and optional ash drawer. You can do this metal work yourself or find a local metal worker to help.
—OR—
» **Firespeaking Barrel Oven Kit** includes steel barrel with insulated door and oven racks ($430), and optionally, firebox with door, ash grate, and ash drawer ($860), from firespeaking.com/barrel-oven-kit

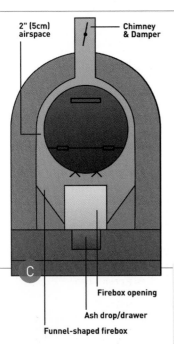

2" (5cm) airspace — Chimney & Damper

Firebox opening

Ash drop/drawer

Funnel-shaped firebox

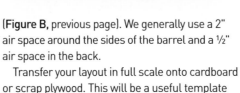

F

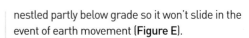

G

D

E

(**Figure B,** previous page). We generally use a 2" air space around the sides of the barrel and a ½" air space in the back.

Transfer your layout in full scale onto cardboard or scrap plywood. This will be a useful template throughout construction. Mark the dimensions and locations of the foundation, base pad, stone or brick walls, ash drawer, firebox door, chimney, and details like space for your plaster.

Draw a vertical cross-section (**Figure C**); it helps anticipate materials needs and the roof height.

4. Prepare your barrel

The minimum amount of metal work necessary for a barrel oven is to fashion at least one rack on which to place baked goods, and a door with a latch that provides a tight seal to keep the heat in (**Figure D**). For the racks, we typically use angle iron and 2"×2" welded wire grid. This work can be done using welding gear, or using simple hand tools including a drill, with nuts and bolts for attaching metal pieces. Or you can use our kit (see Materials list) and let us do the work.

5. Foundation and pad

Prepare a level, well-drained site. Draw the perimeter of your template on the ground, then dig it out, about 12" deep, to reach compacted subsoil. Fill it in with drain rock up to 3"–5" below grade.

We recommend building a "pad" that provides connection and continuity at the base of the oven. You can pour concrete or explore alternatives such as stone, "urbanite" (recycled concrete), or other repurposed materials. An existing concrete slab, patio, or driveway can also serve as a pad.

The pad should be made of units that are as wide as possible (or of one piece). This will unify the load of the smaller bricks that you'll place above. Ensure that your pad is square, level, and

nestled partly below grade so it won't slide in the event of earth movement (**Figure E**).

6. Ash drawer and grate

Lay the first layer of masonry around a long, relatively narrow cavity as shown in **Figure F**. This allows you to collect the ash that accumulates from firing your oven. Our kit includes an ash drawer that fits the space snugly and doubles as an air register. If you're building with adobe brick, use concrete block or fired brick for this first layer to protect the adobe above from ground moisture.

Across the ash cavity, place a metal grate with ¼"–⅜" openings. You can fabricate this from steel plate or from ½" rebar, welded or tied with wire.

7. Firebox and outside walls

To create the funnel-shaped firebox, first build the outside walls (**Figure G**) to rest the diagonally placed bricks against. This shape helps keep the fire organized, burning well, and concentrated over the grate, and begins the sculptural form that takes the heated gases up and around the barrel.

Mortar all points of contact and fill the triangular spaces behind each brick with sand, clay and sand, rocks and clay, etc.

8. Firebox door and lintel

Plan for the top of a course of brick to coincide with the top of your firebox. This will enable you to place a lintel across the firebox opening at the right height. We generally make a lintel across the

> **TIP:**
> METAL EXPANDS MORE THAN MASONRY WHEN HEATED. MORTAR THE LINTEL BRICKS ONLY TO THEMSELVES (NOT TO THE ANGLE IRON) AND BEFORE THE MORTAR SETS, TAP THE ANGLE IRON SO THAT IT MOVES ¼"–½" IN EITHER DIRECTION, TO CREATE AN EXPANSION GAP.

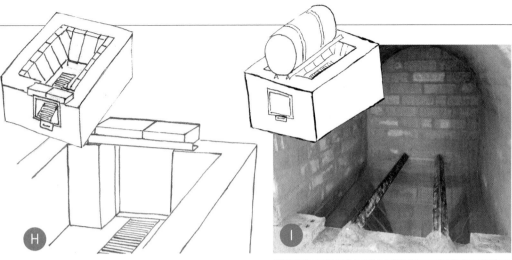

top of the firebox door with a length of steel "angle iron," then place a single course of brick on the lintel to span across (**Figure H**).

9. Barrel supports

Place 2 lengths of angle iron, pipe, or other strong metal across the firebox, from the lintel to the rear wall (**Figure I**). Make sure they're centered and level — shim them with pieces of brick or rock if necessary — then mortar them in place.

Test-fit the barrel on the supports and check that the cooking shelves are level in all directions. You don't want to make slanted birthday cakes!

10. Build the vault

There are a few ways to support the bricks of the vault while you lay them. You can make an arch-shaped armature from metal rod covered with diamond mesh and leave it in place, but we suspect this may contribute to expansion cracks in the plaster later on. Or you could make a rustic arch form out of flexible branches or saplings, and just burn it out on the initial firings of the oven.

We like a removable wooden form for the arch. Cut 2 plywood semicircles with a radius 2" (5cm) greater than the radius of your barrel. Connect these 2 faces with boards the same length as your barrel (**Figure J**). Cover the form with a membrane of lath, mesh, melamine, or cloth to create a solid skin that assists in laying the brick.

Place the arch form so that its top corresponds to the height of the barrel sitting on its supports plus the additional 2" (5cm) of air space above (**Figure K**). Lay your masonry to cover the arch form (leaving room for the chimney), and complete the back wall of the oven. **Figure L** shows a completed vault (after plastering).

11. Chimney

Locate the chimney at the top of the oven, centered along the length of the barrel.

Materials, continued

FOUNDATION & PAD
» **Gravel, drain rock, or fill rock,** about 40gal
» **Cinder blocks, stone, and/or reclaimed concrete** to cover roughly 4'×4' area
» **Solid bricks, 8"×4"×2½"** (10–16) for ash drawer area
» **Mortar mix, 80lb (optional)** —OR—
» **Concrete mix, 90lb (4) (optional)** for a poured pad
» **Steel rebar, ½"×20' (2) (optional)** for a poured pad
» **Lumber, 1×4, 8' long (2) (optional)** for form boards, for a poured pad

OVEN BODY
» **Stone or brick** You can use adobe, compressed earth, or traditional fired brick. Plan on about 300 standard bricks, depending on your design.
» **Clay, ½ cubic yard** for mortar
» **Sand, ½ cubic yard** for mortar
» **Scrap plywood and boards** to build a removable arch form

LINTEL & BARREL INSTALLATION
» **Angle iron, 3/16"×2", 17"**
» **Angle iron, 3/16"×1½", 10'**
» **Ceramic wool**

CHIMNEY
» **Stove pipe, 6" dia. × 5'–6' length**
» **Chimney top, 6"**
» **Cast-iron damper, 6"**
» **Chimney flashing**
» **Storm collar**
» **Trim plate**
» **Stove paint, black**

EARTHEN PLASTER
» **Local clay, red or brown,** 20gal
» **Sand, 50gal**
» **Chopped straw, 15gal**

M

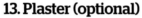

N

» **If you used a metal armature**, cut a hole in the mesh, then attach the chimney to the armature and lay the final bricks around it.

» **If you used a removable arch form**, you'll need to notch the final bricks on top of the oven so that the chimney has a stable seat. Carve them with an angle grinder, then mortar the bricks and chimney securely in place (**Figure M** and **Figure N**).

12. Set the barrel

Let your masonry set at least overnight, then remove the arch form and place the barrel. We place a 4"-wide strip of ceramic wool insulation between the barrel and the seal that surrounds it. Create this seal by filling in the space between the barrel and the arch with cob or carefully placed brick pieces joined with mortar (**Figure O**). A stiff mix made of 1 part clay and 3 parts sand will help to prevent cracking.

13. Plaster (optional)

We plaster our ovens with an earthen plaster. Cement and lime-based plasters are also options. The color for your earthen plaster will come from the beautiful natural colors of the clay and sand you use. Mineral pigments such as oxides and ochre can also be added. Below is a basic recipe.

Mix well and apply with your hands or a trowel (**Figure P** and **Figure Q**). To ensure good adhesion, wet the oven surface thoroughly before applying

EARTHEN PLASTER RECIPE

» 1 part clay or clay-rich subsoil
» 2–3 parts sand
» ½–1 part fiber (short chopped straw or manure)
Mix well and apply with your hands or a trowel.

R

Q

O

P

plaster. We find that 2 coats or more work best.

Tiles and other decorations can be embedded in the wet plaster (**Figure R**).

14. Cook!

Make a few small fires over a week or so to season your oven and prevent any initial shocks. Now you're ready to cook.

You can bake, roast, toast, warm, and dehydrate in a Barrel Oven. Imagine your oven filled with 4 cookie sheets at a time, pizza stones with bubbling pies for your party, stockpots filled with soups and stews, and cast iron skillets baking your cornbread and frittatas to perfection. Casseroles and Dutch ovens are well suited to the barrel oven for long, slow cooking.

The Barrel Oven is hot and ready to bake (350°F–400°F) within 15–20 minutes of getting that good full fire going in the firebox. It can get to 500°–700°F by making a hot, blazing fire and maintaining it — ideal for pizzas, which cook best hot and fast (**Figure S** and **Figure T**).

Once your oven reaches the desired temperature, you can keep a much smaller fire going to maintain the heat as you bake. Use an oven thermometer and experiment!

The Barrel Oven is a very simple pattern that can be modified and varied to improve the cooking experience. What we've presented here is just an overview. For detailed construction tips and materials lists, FAQs, and troubleshooting tips, improvements suggested by other barrel oven builders, and table-tested barrel oven recipes, pick up our book *Build Your Own Barrel Oven*, available in print or PDF ($10) from handprintpress.com/barrel-oven. ✪

See more build photos and share your tips at makezine.com/wood-fired-barrel-oven

Share it: *#makeprojects*

Tools

- » **Wheelbarrow**
- » **Buckets (3–5)**
- » **Shovels (2)**
- » **Mortar board, pan, or tub**
- » **Machete or brick hammer**
- » **Levels (2–3)** We use a 2', a 4', and a small 1' speed level.
- » **Sledgehammer, small**
- » **Cold chisel**
- » **Trowels, diamond shaped (2–3)** for cement and mortar work
- » **Hacksaw**
- » **4½" angle grinder** with masonry (diamond) and metal cutting disks
- » **Eye and ear protection**
- » **Sponges, cleaning brushes, rags**
- » **Welding equipment (optional but recommended)** if you're fabricating your own doors, racks, and drawers. You can drill and bolt these pieces if you don't wish to weld.

EXCERPTED FROM:

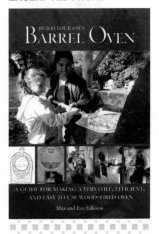

MORE OUTDOOR COOKING
from makeprojects.com

NELLIE BLY SMOKER
Make a hot-or-cold food smoker from a 55gal steel drum.

WOOD GAS CAMP STOVE
Make a simple tin-can stove that runs for free and even sequesters carbon as you cook.

SUN TRACKING PLATFORM
This solar-powered turntable follows the sun, maximizing your solar cooker's potential.

YAKITORI GRILL
Build a Japanese-style grill with ingenious "double-crook" skewers.

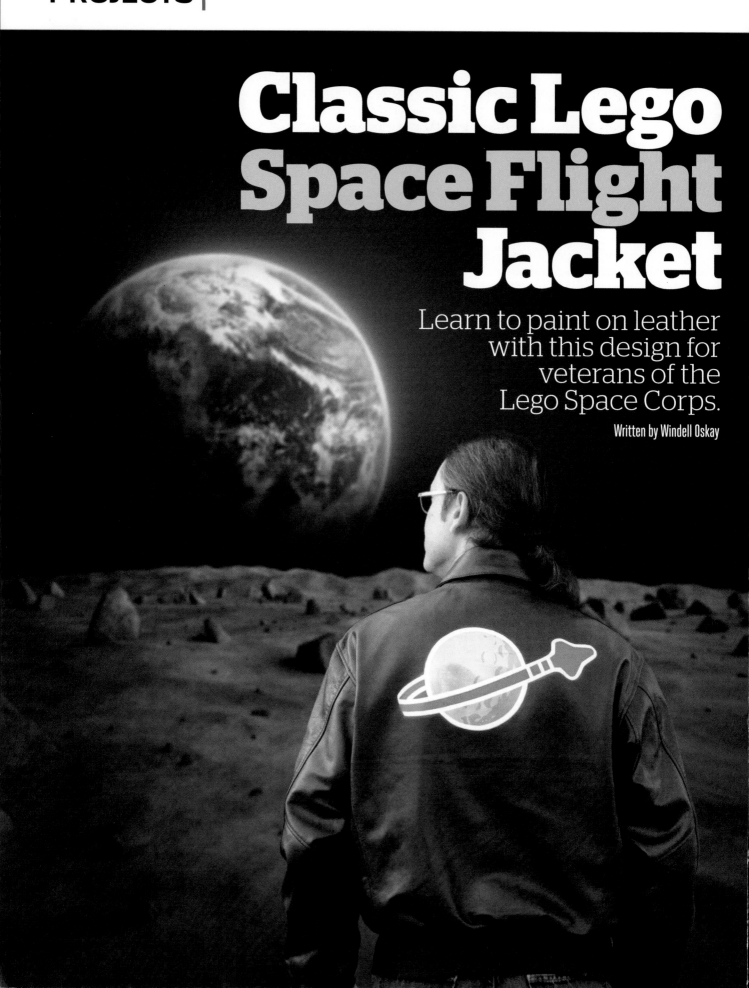

Classic Lego Space Flight Jacket

Learn to paint on leather with this design for veterans of the Lego Space Corps.

Written by Windell Oskay

HERE'S A LITTLE PROJECT THAT WE'VE BEEN WORKING ON FOR A LONG TIME: a custom-painted leather flight jacket featuring the "Classic Lego Space" logo. (Yes, I totally spent years serving in the Lego Space Corps!) If you've ever wanted to make your own painted leather jacket — whatever the theme — here's how to do it.

For this project we used Angelus leather paints, which are highly flexible acrylic paints from the Angelus Shoe Polish Company, designed specifically for painting items like shoes and jackets. We used black, metallic gold, white, and red paints, which are available in 1 oz and 4 oz (and sometimes larger) bottles. We ended up using less than ½ oz of each paint color in this project.

Besides the paints, you willl also need their "Leather Preparer and Deglazer" (for which some people simply substitute acetone), as well as a clear acrylic top-coat finisher. We also used a couple of different sized natural-hair paintbrushes.

The jacket is a "black current issue A-2" leather flight jacket from US Wings, an updated version of the iconic A-2 jacket famously used by pilots in World War II and well known as the classic substrate for painted bomber jackets. We felt black best fit our outer-space theme. Look for one with a solid, one-piece back, so that there aren't any seam lines where you'll be painting.

It's certainly possible to start with a lower-cost jacket or to get a used jacket for much less. However we are following a piece of advice that we learned from our friends in the art car community: If you're going to invest the time in decorating a car, it's a good idea to start with a new one so it lasts longer and you have fewer maintenance issues.

This design is the "Classic Lego Space" logo, no longer in use but prominently featured in popular Lego sets like the Galaxy Explorer and on space-themed minifigures from 1978 to 1987.

If you look closely at pictures of the Galaxy Explorer and minifigures, you might notice that there are actually two versions of the graphic from that era. One, found on larger bricks and flags, has a light-yellow moon with darker craters; while the other, found on small bricks and minifigures, has a metallic gold circle without craters. Since the design on the jacket is large, we went with the cratered version, but exercised artistic license to use light and dark metallic golds, rather than yellows, for the moon's colors.

1. Prepare the stencils

Create the design that you would like to apply to the jacket. For each color in your design, cut a stencil out of cardstock. You can use simple tools like a pencil and a hobby knife for this step, or take more of a high-tech approach. For our design, we used high-resolution pictures of Lego bricks to model the logo in Inkscape (inkscape.org), and laser-cut the cardstock stencils. You may find it helpful to use your cardstock cutouts as mock ups to see how they will look on the jacket in a variety of sizes and positions.

2. Prepare the substrate

Position the stencil for the base (lowest) layer of your painting where you want it to go on the jacket and gently fix it in place with blue painter's tape. Place a board under the jacket back so that you have a smooth and flat workspace.

Use the leather preparer and deglazer solution to strip the shiny outer finish off of the leather inside the stencil. Doing a thorough job of this takes about 30 minutes of hard scrubbing with a washcloth, leaving the cloth thoroughly blackened and that area of the jacket still black, but with a matte finish.

**Time Required:
2 Days
Cost:
$100-$400**

WINDELL OSKAY
is the co-founder of Evil Mad Scientist Laboratories. He has one childhood's worth of experience building Lego spaceships.

Materials

- » **Leather jacket, black, A-2 style**
- » **Leather paints, 1 oz in each color that you will use, and clear top-coat finisher** Angelus brand, angelusshoepolish.com
- » **Leather preparer and deglazer** angelusshoepolish.com
- » **Cardstock**

Tools

- » **Laser cutter or hobby knife**
- » **Computer and Inkscape software**
- » **Natural-hair paintbrushes**
- » **Painter's tape**

CAUTION: WEAR GLOVES FOR THIS STEP — YOU PROBABLY DON'T NEED TO STRIP THE FINISH OFF YOUR FINGERS.

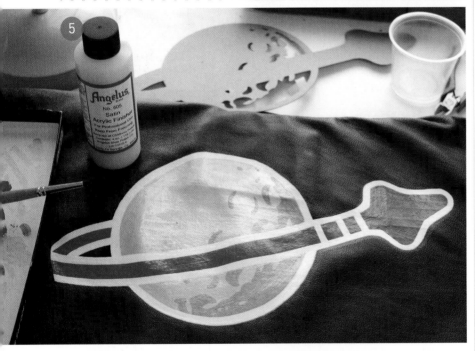

3. Paint a base layer

After the leather dries from the prep stage, it's time to begin painting. Angelus recommends making the first layer thin and allowing it to dry thoroughly before adding the subsequent 2-3 layers necessary to reach full opacity. We used white for our base layer, even under the parts that will be red or gold in the final painting. The brush strokes are quite visible and the opacity is marginal with only a single layer.

This is the easiest layer of painting, because — with the exception of black areas like our little triangle — you can just paint over all of the areas that have that initial matte appearance.

4. Add color and opacity

Once the paint is dry (30 minutes, no touching), use your other stencils as guides to mark where the color layers will go. Align each stencil to the base layer and trace it very gently with a pencil or spudger. Using those (barely visible) marks as guides, paint each color in your design, in sequence. For each color, let the paint dry for at least 30 minutes before doing the next layer, and alternate between all of the colors in your design.

After four layers of white and three each in red and gold, our white and colored areas were well defined, bright, and distinct.

NOTE: PAINTING THE INTERIOR CURVES TO APPEAR SMOOTH TAKES A FAIR AMOUNT OF TIME AND STEADY HANDS.

5. Mixing custom colors

Our design called for light and dark metallic gold. However, Angelus only offers a single (but gorgeous) shade of metallic gold paint, which is alternately referred to as "Antique Gold" or just "Gold." The good news is that you can custom mix your own paint colors. To make a darker gold paint, simply mix in a few drops of black.

For low-contrast accents like our craters — dark gold atop regular gold — a single layer of additional stencil-guided paint works well. Touch it up, as needed using both the lower and upper paint colors..

6. Apply clear finish

Allow the paints to dry overnight, and add the final top coat — a thin and colorless acrylic finish that dries clear. The finisher is available in five different luster levels, from matte to glossy. We used "Satin," but found it to be glossier than the rest of the jacket, so we'd recommend "Flat Top," listed as having the lowest sheen. Apply 2 coats of the finish and allow it to dry overnight.

And we're done!

A final word about the paints: These paints are (as advertised) incredibly flexible and well matched to leather. You can flex and fold the material, and the paint goes right along with it, flexing and folding as though it isn't there at all. Neat stuff! ●

See more pictures and share your comments at makezine.com/lego-space-jacket
Share it: #makeprojects

Sing-a-Long Song Devocalizer

Cancel out vocals, isolate instruments, and create fascinating audio effects with this inexpensive DIY device.

By Jeffrey M. Goller and Nathan Goller-Deitsch

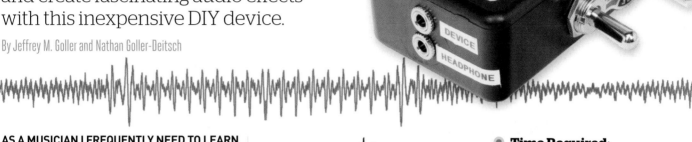

AS A MUSICIAN I FREQUENTLY NEED TO LEARN NEW SONGS, and it can be difficult to hear individual instruments in recordings I want to study. I have often wished for an easy way to eliminate or reduce the vocals and isolate the instruments. Expensive, inconvenient solutions exist, but I wanted a cheap method that I could use anywhere, on any device with a ⅛" stereo headphone jack.

Then, one night, I was fiddling with a pair of headphones with a defective plug. When moved a certain way, it gave the exact vocals-canceling effect I had been looking for!

To understand what happened, I researched online, bought some components at RadioShack, and tested various wiring combinations. Together with my 10-year-old son Nathan, I designed and built a circuit that replicates the effect.

How It Works

When songs are mixed from individual tracks, the waveforms of the isolated instruments and voices are added together to form the left and right channels of the stereo mix. Typically, vocal tracks are placed in the center of the mix, which results in mathematically identical waveforms being sent to the left and right channels.

The Devocalizer lifts the common ground for the stereo signals, and instead uses the opposite channel as the new ground, putting the signals 180° out of phase from each other. Each stereo channel is added to an inverse copy of the opposite channel, canceling any audio that's mixed in both channels. Result: Vocals magically vanish!

The isolation of instruments using the technique can be remarkable. We've found that about 20% of songs have fantastic vocal cancellation — you could use them for karaoke, no problem. In 50%–60% of songs, the effect is less pronounced

The waveforms of various musical instruments and voices are combined to form the left and right channels of a stereo mix.

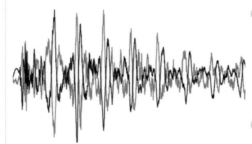

The Sing-a-Long Song Devocalizer causes the stereo signals to become 180° out of phase.

Result: Anything mixed evenly to both right and left channels will magically disappear.

but you'll still hear instrumental parts you never heard before. About 15% are duds, and 5% have a fascinating, totally unexpected, robotic audio sound that's difficult to describe. Experiment, have fun, and let us know if you find any tunes with especially pronounced or weird effects! ◗

Time Required:
1 Hour
Cost:
$40-$60

JEFFREY M. GOLLER
is an emergency physician in sunny Charleston, S.C., who originally hails from Cincinnati. He plays the upright bass with The SouthRail Bluegrass Band, and enjoys designing and building musical instruments.

NATHAN GOLLER-DEITSCH
is a 10-year-old home-schooled programmer/developer (who also happens to be the son of Jeffrey Goller). He programs in Visual Basic .NET, PHP, and HTML, and wants to move to Silicon Valley (when he is old enough) to have his own start-up company.

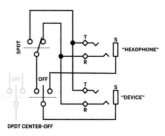

For step-by-step build notes and video, visit makezine.com/song-devocalizer

Share it: #makeprojects

Gunther Kirsch

Open-Source CNC Furniture

Forget Ikea. Fabricate your own stylish, personally customized furniture on a CNC machine near you.

Written by Anna Kaziunas France

OPEN-SOURCE HARDWARE (OSHW) ISN'T JUST MACHINES AND ELECTRONICS — with the rise in popularity of CNC routers and laser cutters, OSHW has expanded into furniture. The result is a fabrication movement where designs are shared globally but fabricated locally, and parametric designs can be infinitely configured for personal fabrication.

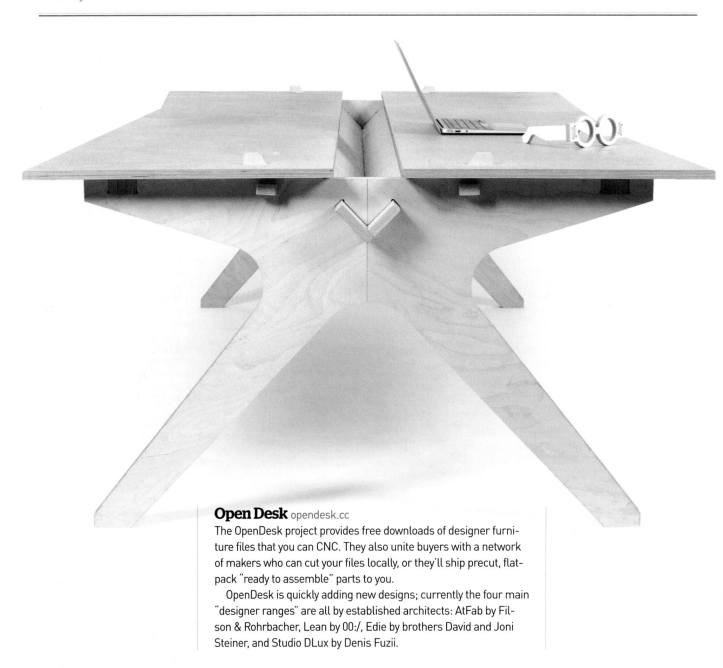

Open Desk opendesk.cc
The OpenDesk project provides free downloads of designer furniture files that you can CNC. They also unite buyers with a network of makers who can cut your files locally, or they'll ship precut, flat-pack "ready to assemble" parts to you.

OpenDesk is quickly adding new designs; currently the four main "designer ranges" are all by established architects: AtFab by Filson & Rohrbacher, Lean by 00:/, Edie by brothers David and Joni Steiner, and Studio DLux by Denis Fuzii.

SketchChair

sketchchair.cc

Released in 2011 after a successful Kickstarter, SketchChair's open software makes it easy to design functional, customized furnishings using a 2D "sketching" interface that turns your drawings into chairs. Files can then be exported and CNC routed, laser cut, or made in miniature on a paper cutter.

The Layer Chair

makezine.com/layerchair

Jens Dyvik created his open-source Layer Chair while exploring how to work with large organic surfaces on the ShopBot. Using Grasshopper and Rhino CAD software, he created a definition that takes the input from two adjustable curves and outputs machinable files with assembly alignment guides. When assembled with dowels and glue the 2D slices create a "stair-stepped" 3D chair surface.

As intended, this open-source chair has mutated into many different versions, including rustic formal dining chairs in the shape of mountain peaks in Norway, a tall stool in New Zealand, an Amsterdam Edition black table and chairs, and a custom performance chair for world famous cellist Frances-Marie Uitti.

Ronen Kadushin

ronen-kadushin.com

Ronen Kadushin has been releasing open designs since 2004. His Italic Shelf system has two basic parts that can be assembled in many configurations. Using Kadushin's "Controlled Collapse" structural position locking, the shelf maintains its strength and stability, even when stacked high.

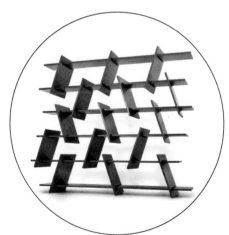

AtFab

atfab.co

Architects Anne Filson and Gary Rohrbacher are the creators of AtFab, a designer line of commercial CNC furniture whose six designs are also downloadable from OpenDesk. Yet they have higher ambitions — they're currently creating fully parametric furnishings whose dimensions, details, material thickness, and slot size can be easily transformed and fabricated.

As they expand their catalog, they've pledged to offer the files to the maker community before they are available for retail. You can't yet export the DXFs, but you can try out the early release of their browser-based Processing apps at filson-rohrbacher.com/portfolio/ autoprogettazione. (See sidebar for more details).

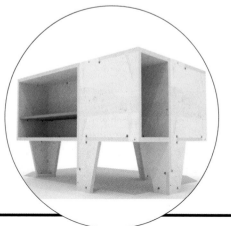

WHERE TO FABRICATE?

Find a CNC router or an independent fabricator near you: makezine.com/ where-to-get-digital-fabrication-tool-access

ANNA KAZIUNAS FRANCE
is the digital fabrication editor at MAKE. She's also the dean of students for the Global Fab Academy program, the co-author of *Getting Started with MakerBot*, and the editor of the book *Make: 3D Printing*.

You Had Me at Open Source + Parametric

As a lover of all things customizable, I simply had to build a custom AtFab design, but the online apps couldn't yet spit out DXFs. I inquired about an alpha version and received their One to Several Table parametric configurator for Processing. I used it to fabricate custom standing-height workbenches for the new MAKE Providence, R.I., office. AtFab has graciously agreed to allow MAKE to distribute the Processing app — grab it and **make your own custom CNC Maker Bench** a makezine.com/ cnc-maker-bench.

How Many Days Until the Next MAKE?

Use the Intel Galileo to make networked gadgets — like this countdown display for your favorite magazine. Written and photographed by Matt Richardson

USING THE INTEL GALILEO FOR INTERNET-OF-THINGS PROJECTS is a natural fit. With Arduino pin compatibility, networking, and Linux, it's a powerful yet flexible board that can control hardware, handle computing jobs ordinary microcontrollers can't, and send and receive information from the internet. And what information is more important than knowing when to expect the next volume of MAKE?

I've created a website called How Many Days Until MAKE Comes Out? that serves one simple purpose: It tells you how many days until the next issue of MAKE magazine is scheduled to hit newsstands. The source code is available at github. com/mrichardson23/nextmakemagazine if you want to see how I made it.

Go to nextmakemagazine.appspot.com in your web browser, and you'll see the information is formatted to be viewed and understood by a human. Next, visit nextmakemagazine.appspot.com/simple and you'll see the server is also configured to speak directly to microcontrollers, by stripping away all the extra style and language and only returning the number of hours until the next issue is released.

Your Galileo can use its internet connection via Ethernet to connect to this URL, receive the data, and evaluate how to display it in your home. We'll pass this data from the Linux side to the Arduino side — a powerful feature of this board — and then display it on a standard LCD.

NOTE: FOR THIS PROJECT, YOU'LL NEED TO BOOT OFF OF GALILEO'S SD CARD IMAGE, WHICH YOU CAN DOWNLOAD FROM INTEL AT makezine.com/go/galileoimage.

This project is adapted from the new book *Getting Started with Intel Galileo*, available from the Maker Shed (makershed.com).

Testing the Connections

First, make sure your Galileo can connect to the server.

1. Connect the Galileo via an Ethernet cable to your network, plugging it into your router or an active Ethernet jack.
2. Connect the Galileo to power.
3. Connect your computer to Galileo via the USB client port.
4. Launch the Arduino IDE software and select File → Examples → Ethernet → WebClient.
5. Click Upload.
6. Open the Serial Monitor.

You'll see text start to appear in the Serial Monitor. This example programs your Galileo to do a Google search for the term "Arduino." As the

WHAT IS GALILEO?

The **Intel Galileo** is an innovative new microcontroller that runs Linux out of the box and supports Arduino programming and most Arduino shields. It's based on the Intel Quark SoC X1000, a 32-bit Intel Pentium-class system on a chip, so it's more capable than many other controller boards.

In addition to the familiar Arduino hardware, the Galileo board has a full-sized Mini-PCI Express slot, 100Mb Ethernet port, MicroSD slot, RS-232 serial port, USB Host port, USB Client port, and 8MByte NOR flash memory.

The Galileo also has the ability (unlike others) to multitask while operating an Arduino sketch — which opens up a world of new opportunities for your projects.

HTML response from Google's server is received, it sends those characters to the Serial Monitor.

Now that you're sure the network connection is working, tell your Galileo to connect to the How Many Days Until MAKE Comes Out? server:

1. Create a new sketch and enter the code in **Figure A**.
2. Upload the code and then open the Serial Monitor.

If it worked, you should see the number of hours out printed every 5 seconds in the Serial Monitor.

The `loop` function has only 2 lines of code. The `delay(5000)` is what ensures that each iteration of the loop only happens every 5 seconds. But what about `Serial.println(getHours());`? The innermost function, `getHours()`, is defined right below the `loop` function. It requests data from the server, stores that response in a file, and then reads the file and returns an integer value representing the number of hours you'll need to wait for the new magazine. It's the `atoi()` function that looks at the ASCII characters sent by the server, (say, 4 and 5) and outputs their value as an integer (45), which you can use for arithmetic.

Having the Linux command write the data to a file and then having the Arduino sketch read that file is just one way that you can get data into your sketch. This unique Galileo feature is a powerful way to connect different parts of a project together since there are so many ways to read and write files.

Parsing JSON with Python

The example in **Figure A** is easy because it handles only one simple piece of data. But lots of web services provide several pieces of data structured in a format called *JSON*, or JavaScript Object Notation.

JSON has become the standard format for transmitting structured data through the web. If you want to read data from a site that offers JSON, you'll have to parse it. Since this would be difficult to do with Arduino code, you can use other languages on the Galileo to do this job and pass the appropriate information into the Arduino code.

To preview JSON data, visit nextmakemagazine.appspot.com/json in your web browser:

```
{"totalHours":1469.0,"volumeNumber":"40","daysAway":61}
```

There are 3 key/value pairs: the number of hours until the next issue, the next volume number, and the number of days until the next issue.

The code in **Figure B** uses the Python programming language to connect to the server's JSON feed at nextmakemagazine.appspot.com/json and parses the volume number and number of hours.

1. Connect to Galileo's command line using SSH, Telnet, or serial.
2. Change to root's home directory.
`cd /home/root/`
3. Launch the text editor vi with the filename json-parse.py to create that file.
`vi json-parse.py`
4. Along the left side of the screen you'll see a column of tildes (~). Type the letter `i` to enter insert mode. An I will appear in the lower left corner of your screen.

**Time Required:
1-2 Hours
Cost:
$80-$100**

MATT RICHARDSON
MAKE contributing editor Matt Richardson (mattrichardson.com) is a Brooklyn-based creative technologist and Resident Research Fellow at ITP.

Materials
» **Intel Galileo microcontroller board** Maker Shed item #MK-ING01, makershed.com
» **Ethernet cable**
» **USB cable**
» **LCD character display,** 16×2 Maker Shed #MKAD30
» **Breadboard** Maker Shed #MKEL3, MKKN3, or MKKN2
» **Jumper wires** Maker Shed #MKSEEED3
» **Potentiometer, 10kΩ or 2kΩ**
» **Header pins, male (optional)** if your LCD lacks breadboard-compatible pins

Tools
» **Computer with Arduino IDE software** free download from arduino.cc
» **Soldering iron and solder (optional)** if your LCD lacks breadboard pins

```
A
void setup() {
}
void loop() {
Serial.println(getHours());
delay(5000);
}
int getHours() {
char output[5];
system("curl http://nextmakemagazine.appspot.com/simple >
response.txt");
FILE *fp;
fp = fopen("response.txt", "r");
fgets(output, 5, fp);
fclose(fp);
return atoi(output);
}
```

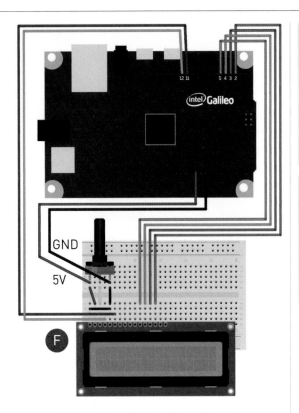

5. Enter the code from **Figure B** into vi.

6. Hit the Escape key to switch from insert mode to command mode. The **I** in the lower left corner will disappear and you'll see a dash instead.

7. Type **:x** and press Enter to save the file and exit vi.

8. Test the script by executing the code from the command line.

```
# python json-parse.py
```

If you got everything right, you should see the following output on the command line:

```
Volume 40 will be released in
1473.0 hours.
```

As you can see, parsing a JSON response from a website isn't hard when you have Python available to you on Galileo. Now you'll simply connect the response from Python to your Arduino code.

To try that now, first modify *json-parse.py*:

1. On Galileo's command line, be sure you're still in root's home directory:

```
# cd /home/root/
```

2. Open the file for editing in vi:

```
# vi json-parse.py
```

3. Type the letter **i** to enter insert mode. An **I** will appear in the lower left corner of your screen.

4. Edit the file so that it reflects the code in **Figure C**.

5. In the Arduino IDE, create a new sketch with

the code in **Figure D**. You'll see it's very similar to the Arduino code **Figure A**. Instead of calling `curl` from the command line, it uses Python to run the script you wrote.

6. Upload the code to the board and open the Serial Monitor.

Now you should see the response from the server as the number of days until the next issue of MAKE comes out.

Connecting an LCD Character Display

What good is this information if it can only be seen in your Serial Monitor? Let's hook up an LCD display to read out the info where anyone can see it (**Figure E**).

To connect the LCD to Galileo:

1. Disconnect your Galileo board from your computer's USB port and from power.

2. Insert the LCD into the breadboard (solder header pins onto it if necessary).

3. Insert the potentiometer into the breadboard as well.

4. Using your jumper wires, connect the potentiometer and LCD to Galileo as shown in **Figure F**.

5. Connect power to Galileo.

6. Connect Galileo to your computer via USB.

7. From the Arduino IDE, upload the code in **Figure G**.

Conclusion

Now you can easily keep tabs on when the next MAKE will hit newsstands! This is a simple example — you can just imagine the tricks Galileo can do with almost any data on the web. Pick up a copy of *Getting Started with Intel Galileo* for more network-connected Galileo projects. ◆

B
```
import json
import urllib2

httpResponse = urllib2.urlopen('http://↵
nextmakemagazine.appspot.com/json')
jsonString = httpResponse.read()

jsonData = json.loads(jsonString)

print "Volume", jsonData['volumeNumber'], "will be↵
released in", jsonData['totalHours'], "hours."
```

C
```
import json
import urllib2

httpResponse = urllib2.urlopen('http://↵
nextmakemagazine.appspot.com/json')
jsonString = httpResponse.read()

jsonData = json.loads(jsonString)
print jsonData['daysAway']
```

D
```
void setup() {
}
void loop() {
Serial.println(getDays());
delay(5000);
}
int getDays() {
char output[5];
system("python /home/root/json-
parse.py > /response.txt");
FILE *fp;
fp = fopen("response.txt", "r");
fgets(output, 5, fp);
fclose(fp);
return atoi(output);
}
```

G
```
#include <LiquidCrystal.h>
LiquidCrystal lcd(12, 11, 5, 4,
3, 2);
void setup() {
lcd.init(1,12,255,11,5,4,3,2,0
,0,0,0);
lcd.begin(16, 2);
lcd.setCursor(3, 0);
lcd.print("days until");
lcd.setCursor(0, 1);
lcd.print("MAKE is here!");
}
void loop() {
lcd.setCursor(0, 0);
lcd.print(" ");
lcd.setCursor(0, 0);
lcd.print(getDays());
delay(30*60*1000);
}
int getDays() {
char output[5];
system("python /home/root/json-
parse.py > /response.txt");
FILE *fp;
fp = fopen("response.txt", "r");
fgets(output, 5, fp);
fclose(fp);
return atoi(output);
}
```

Download the project code and share your Galileo ideas at makezine.com/galileo-make-countdown

Share it: *#makeprojects*

Duct Tape Double

Written by Paloma Fautley ■ Illustrations by Julie West ■ Photos by Gunther Kirsch

THE DUCT TAPE DOUBLE PROJECT IS A QUICK AND EASY WAY TO MAKE A CUSTOM MANNEQUIN. It uses cheap materials you can find around the house to create a dimensionally accurate replica. Use your new double to craft custom garments that fit perfectly every time, or step it up a notch and venture into well-tailored wearable technology.

1. Prepare the model
Use a tight-fitting shirt that covers the area that you want to replicate. If you want to add length beyond the shirt, cover the additional area with plastic wrap. Make sure that the model is comfortable and can breathe easily.

2. Wrap it up!
Wrap the model with duct tape. Use the patterns provided as a general guideline. Don't pull too tightly or the double won't accurately represent your true size.

Cover every desired area with 2–3 layers of duct tape so it is nice and sturdy.

3. Remove and stuff
Cut the duct tape double off your model carefully, slicing up the back and making sure to not cut the model in the process. Carefully slide the double off. Be gentle and try not to lose any features while doing so.

Stuff the double with old clothes, newspaper, or other materials to make it sturdy. Do not overstuff or you will lose the shape.

Tape up the openings, and you're finished! ⊘

For a video of the build and detailed drawings of the wrapping patterns, visit makezine.com/duct-tape-double
Share it: *#makeprojects*

PALOMA FAUTLEY is an engineering intern at MAKE. She is currently pursuing a degree in robotics engineering and has a range of interests, from baking to pyrotechnics.

You will need:
» **Duct tape**
» **Old shirt you don't mind ruining**
» **Newspaper or old clothes for stuffing**
» **Scissors**
» **Plastic wrap (optional)**

Air Rocket Glider

Build a new kind of folding glider that's blasted skyward by a compressed air launcher! Written by Keith Violette with Rick Schertle

KEITH VIOLETTE
is a husband, dad, inventor, maker, mechanical designer at DEKA R&D, and co-founder of AirRocketWorks.com. He has designed prosthetic arms, medical devices, kids' toys, and ultra-precision air bearing devices.

RICK SCHERTLE
(schertle@yahoo.com) teaches middle school in San Jose, Calif., and leads after-school maker clubs. With his wife and kids, he loves all things that fly. Along with Keith, he's the co-founder of AirRocketWorks.com.

Jeffrey Braverman

My Compressed Air Rocket Launcher project in MAKE Volume 15 proved to be wildly popular. We've turned it into a kit for the Maker Shed and launched tens of thousands of rockets at Maker Faire. (The Version 2.0 launcher is now available, see page 83.)

After I wrote a Folding-Wing Glider project in Volume 31, I thought it would be cool to combine the two projects. Then I got an email from MAKE reader Keith Violette, who was a fan of both.

A maker collaboration was born. Now we present to you the world's first Air Rocket Glider.
—Rick Schertle

AFTER READING MAKE'S FOLDING-WING GLIDER ARTICLE, MY KIDS REALLY WANTED TO BUILD ONE

(makezine.com/rocketglider). A few hours later, we were out in the New Hampshire snow, launching the glider. I was intrigued by the cool mechanism that allows the wings to be stowed parallel to the plane body. The wings are held back during launch and ascent by wind resistance, only to pop out once the plane slows near its maximum altitude.

That same day we also happened to be launching our compressed air rockets. It's the "maker way" to figure out fun ways to combine things — and so the Air Rocket Glider (ARG) was invented.

Here's how to build your own Air Rocket Glider. You can also try our new kit (check out airrocketworks.com). Either way, it's a lot of fun.

1. Print the plastic parts

Download the part files from makezine.com/air-rocket-glider and 3D print them in ABS plastic at 100% fill: left fuselage,

right fuselage, wing pivot halves (2), tail fins (3), wing reinforcements (2), and the 2-part nose mold (optional, see Step 2). Sand if needed to correct warping.

2. Cast or print the nose
Bolt the nose mold together with the #10-32 hardware. Mix 20ml of urethane resin, fill the syringe, and expel the air bubble.

Slowly inject the resin into the large hole in the mold, until a puddle forms on top at the small vent hole. Let the resin cure.

OPTIONALLY, you can 3D-print the soft nose in flexible filament at 0% fill. We've provided preconfigured .ini slicing files.

3. Fabricate the body tube
Cut the body tube to 9", using a hacksaw (**Figure 3a**). Sand the ends smooth and perpendicular, and remove any burrs using 220-grit paper (**Figure 3b**).

Mark a line down the length of the body tube with a pencil. I like the old rocket fin trick — press the tube into the corner of a doorjamb, and use the jamb as a guide to draw your line (**Figure 3c**).

Download and print the template from makezine.com/air-rocket-glider, and cut out the fin guide. Wrap it around the tube, align the Vertical Tail mark with your first pencil line, and mark lines for the other 2 fins.

4. Assemble the fuselage
Seat the rear end of the nose in its pocket in the left fuselage. Using ¼" lengths of filament in the corner holes as alignment pins, super-glue and clamp the fuselage halves together, capturing the nose between (**Figure 4a**, following page). Let the glue dry, then clean out the wing pivot hole using a ⅜" drill bit (**Figure 4b**, following page).

Time Required:
3–4 Hours
Cost:
$15–$20

Materials
- » **Nylon tube, ¾" OD, ¹¹⁄₁₆" ID, 9" length** McMaster-Carr #8628K61, mcmaster.com, sold in a 5' length
- » **Spring wire, 0.045"–0.050", 9" length** or piano wire or TIG welding wire
- » **Rubber band, size #16 (2½" ×¹⁄₁₆")** McMaster #12205T74 or Staples #808576
- » **Standard (F1667 STFCC-04) staples (2)**
- » **Balsa wood, ³⁄₃₂" ×3" ×8" (2)**
- » **Super glue, impact resistant** such as Loctite 411
- » **Castable urethane resin, 40 Durometer hardness** such as Smooth-On PMC-724, smooth-on.com. The 1lb trial size will make lots of nose cones. To color the resin, try Smooth-On's So-Strong tints.
- » **Syringe, 50cc** McMaster #7510A665
—OR—
- » **Flexible filament** such as NinjaFlex polyurethane or flexible PLA, if you'd rather print the nose than cast it
- » **Compressed Air Rocket Launcher Kit (version 2.0)** from airrocketworks.com
—OR, IF USING THE OLDER PVC LAUNCHER—
- » **Soft steel wire, 18"** from a coat hanger
- » **PVC pipe, Schedule 80, ³⁄₈" NPT, 12" length, threaded on one end** McMaster #9173K412, for the launch tube
- » **Reducer bushing, ¾" NPT male to ³⁄₈" NPT female** McMaster #4596K405

Tools
- » **3D printer with ABS filament, and (optionally) flexible filament**
- » **Hacksaw**
- » **Sandpaper, medium (150–400 grit)**
- » **Pliers, needlenose**
- » **Scale or tape measure**
- » **Stapler** that hinges open for tacking
- » **Cardboard**
- » **Fine wire**
- » **Drill bit, ³⁄₈"** no drill needed
- » **Paper clip, large**

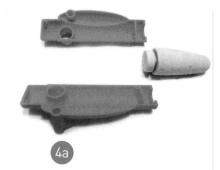

4a

8a

8b

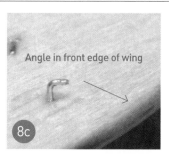

Angle in front edge of wing

8c

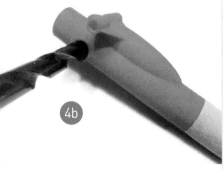

4b

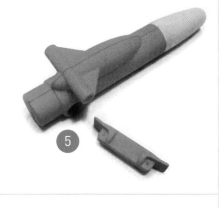

5

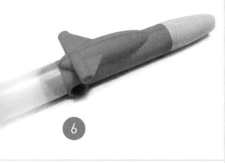

6

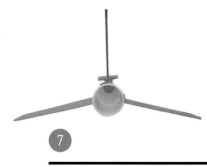

7

5. Assemble the wing pivot

Carefully align the 2 halves of the wing pivot and glue them together. Test-fit the wing pivot in the fuselage and ensure that it rotates freely without binding. Sand down the glue seam or the outer faces of the wing pivot if needed (**Figure 5**).

6. Mount the body tube

Glue the fuselage into the body tube, aligning the fuselage's top seam with the Vertical Tail line on the tube (**Figure 6**). Work quickly — the glue sets up in only a few seconds.

7. Mount the fins

Wrap 220-grit sandpaper around the body tube, then use it to sand the curve into the base of each tail fin to ensure a matching radius and a good surface for adhesion. Wrap a piece of masking tape around the body tube 2" from the open end.

Apply glue to the curved base of one of the fins. Carefully align the fin on your Vertical Tail pencil line, and align its leading edge with the tape edge. Bond the fin in place. Bond the remaining 2 tail fins, aligning them with your marks (**Figure 7**). Remove tape and pencil lines.

8. Build the wings

Cut the balsa wings as shown on the template. Use sandpaper to round the leading (front) and trailing (rear) edges. This will help prevent cracks and improve the aerodynamics.

Apply glue to the top and bottom of the wing, 3/8" out from the base edge. Quickly slide the wing reinforcement onto the wing, aligning its notch with the wing notch (**Figure 8a**).

Now you'll add the staples that will anchor the rubber band. Overlay the wing template on the wing, and open a standard stapler into "tacking" mode. Place the wing on 2 layers of cardboard for a temporary backing, and staple through each wing where indicated. Pull off the template, taking care not to dislodge the staple (**Figure 8b**).

Apply super glue over the crown of the staple on the underside of each wing and let it cure.

Bend the inner staple leg down flat to the wing top, and cover it with super glue. Also glue the base of the outer, standing leg of the staple, and let it cure. Finally, bend a small hook into the standing leg, pointed toward the angle in the leading edge of the wing (**Figure 8c**).

9. Bend the pivot wire

In the middle of the 9" wire, use needlenose pliers to bend a gentle radius that matches the outer diameter of the body tube. The legs of the wire should be parallel, and roughly equal in length — you'll trim them later (**Figure 9a**).

Now grip the wire just below the midline of the tube, as shown, and make an approximate 100° forward bend in each leg of the wire. Again, the legs should be parallel (**Figure 9b**).

10. Assemble the moving parts

Insert the rubber band through the upper hole in the fuselage.

Install the wing pivot in the fuselage, then align the notches in the wings with the holes in the wing pivot. Slide the wing pivot wire through the wing reinforcements, starting at the trailing edge. It should slide along the base of each wing, through the holes in the wing pivot, and out the leading edges of the wings. This may take a couple of tries. Pivot the wings to ensure smooth operation. Trim the excess wire flush to the leading edges.

Stretch the rubber band and hook the ends to the staple hook on each wing.

11. Test your wings

Start with the wings folded back (**Figure 11a**). When released, they'll hinge forward on the pivot wire (**Figure 11b**), and then rotate on the plastic pivot into gliding position (**Figure 11c**). Ensure that they open quickly, evenly, and smoothly. If one side opens faster than the other, equalize the tension in the rubber band on either side.

Now check the angles of the wings in the open position. You can adjust the *angle of attack* by altering the 100° bends in the wing pivot wire.

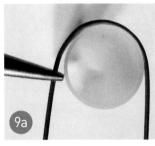

9a

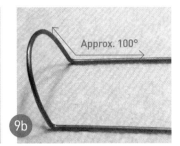

Approx. 100°

9b

Insert the wing pivot wire down through the wing reinforcements and the holes in the wing pivots

10

11a

11b

11c

The *dihedral angle* should be 3° to 6° as built here. You can alter it by adding tape or thin shims to the top of the wing where the wing pivot contacts the wing reinforcement. Greater dihedral angle makes the plane more steady but reduces lift.

12. Balancing and tuning

Due to the varying densities of balsa, it's important to balance your Air Rocket Glider left to right. Invert the plane and roll it side to side on 2 fingertips (**Figure 12**). If one wing is heavier than the other, add bits of packing tape to the tip of the lighter wing until the plane balances. This will help it fly straight and true.

If you're flying your ARG in a smaller field or park, you can purposely weight one wingtip to upset the balance. This will cause the ARG to spiral down to the ground and not drift too far from the launch site.

On windier days, you can add a second rubber band to increase the opening power of the wings. This will cause them to deploy slightly sooner, at a lower altitude, but will help prevent the wind from causing the plane to tumble or spin without opening its wings fully.

Use fresh rubber bands to ensure proper wing operation. During storage, unhook the rubber band to prevent it from stretching.

Your Air Rocket Glider is complete!

13. Launching the Air Rocket Glider

The ARG launches off of a ⅜" NPT pipe. A bent piece of wire holds the wings in their folded position until the glider is launched. The new Compressed Air Rocket Launcher Version 2.0 kit includes this launch tube, adapter, and the wing holder wire.

If you have MAKE's older Compressed Air Rocket launcher, its launch tube is too large — but

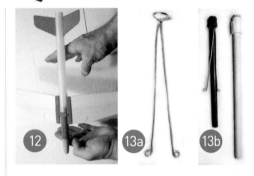

12   13a   13b

that's easily remedied. Remove the existing ½" PVC launch tube and swap it for a ⅜" NPT pipe, 12" long, threaded on one end. Screw the ⅜" pipe into a reducer bushing, ¾" male NPT to ⅜" female NPT, then screw the reducer into your existing sprinkler valve.

Finally, bend the wing holder wire from an 18" length of coat hanger wire and install it on the base of your new ⅜" launch tube, as shown in **Figures 13a, 13b, and 13c**. Ready for launch!

Let us know how your ARG flies, and join the air rocket community at airrocketworks.com. ◢

Download the 3D files and share your build tales at makezine.com/air-rocket-glider
Share it: *#makeprojects*

13c

Flora NeoGeo Watch

Make a styling LED timepiece with GPS navigation and compass modes built in. Written by Becky Stern and Tyler Cooper

Time Required:
A Weekend
Cost:
$140

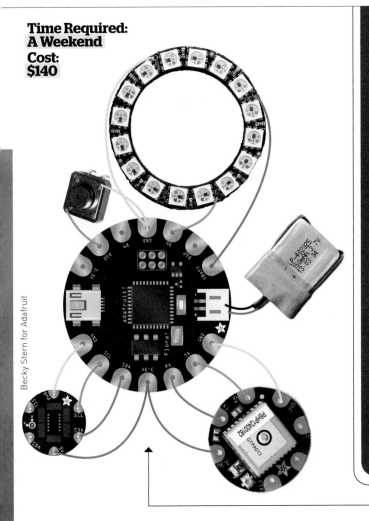

Becky Stern for Adafruit

BECKY STERN (sternlab.org) is a DIY guru and director of wearable electronics at Adafruit. She publishes a new project video every week and hosts a live show on YouTube. She lives in Brooklyn, N.Y., and belongs to art groups Free Art & Technology and Madagascar Institute.

TYLER COOPER
Tyler Cooper is a creative engineer at Adafruit Industries, where he helps develop the Adafruit Learning System. In 2010, he co-founded the open-source hardware company Coobro Labs. He's also co-owner of the Minneapolis/St. Paul, Minn., makerspace Nordeast Makers.

WHY FLORA?

Flora is Adafruit's Arduino-compatible wearable electronics platform. Measuring only 1¾" in diameter, it's small enough to embed into any wearable project, and it has large pads for sewing with conductive thread. The round shape means there are no sharp corners to poke through your garment, and the 14 pads are laid out to make it easy to connect a variety of sensors and modules, such as Flora NeoPixels: addressable, color-changing LED pixels.

With its powerful ATmega32u4 microprocessor, Flora has built-in USB support. It's programmed via Mac or PC with free software you download online. Adafruit publishes hundreds of tutorials and dozens of free code libraries for Arduino-compatible boards, so you'll never be lacking project ideas or sample code to get you started.

Materials

» **Adafruit Flora microcontroller board** Maker Shed #MKAD58, makershed.com or Adafruit #659, adafruit.com
» **Flora Wearable Ultimate GPS Module** Adafruit #1059, or get both the Flora and its GPS module in the Flora GPS Starter Pack, Maker Shed #MKAD51 or Adafruit #1090
» **NeoPixel RGB LED ring, 16 LEDs** Maker Shed #MKAD75 or Adafruit #1463
» **Flora Accelerometer/Compass Sensor** Adafruit #1247
» **Tactile switch** Adafruit #1119
» **Battery, LiPo, 3.7V 150mAh, with charger** Adafruit #1317 and #1304
» **Leather watch cuff** Check out Labyrinth Leather on Etsy.
» **Scrap of fabric**
» **E6000 craft adhesive**
» **Binder clips**
» **Thin-gauge stranded wire**
» **Double-stick foam tape**
» **Black gaffer's tape**

Tools

» **Soldering iron**
» **Solder, rosin core, 60/40**
» **Multimeter**
» **Scissors**
» **Flush snips**
» **Wire cutters / strippers**
» **Pliers**
» **Tweezers (optional)**
» **Helping hand tool (optional)**

USE THE FLORA WEARABLE MICROCONTROLLER AND ITS GPS MODULE TO TELL TIME WITH A STUNNING RING OF PIXELS. A leather cuff holds the circuit and hides the battery. The watch is a bit chunky, but it still looks and feels great on tiny wrists!

The circuit sandwich becomes the face of the watch, and you'll use a tactile switch to select its three modes: timekeeping, compass, and GPS navigation. Customize your waypoint in the provided Arduino sketch, and the LEDs will point the way and then tell you when you're arriving at your destination.

This is an intermediate-level project requiring soldering and some precision crafting. Before you begin, read the following guides at learn. adafruit.com: *Getting Started with Flora, Flora Wearable GPS, Flora Accelerometer,* and *Adafruit NeoPixel Überguide.*

This project is adapted from our book *Getting Started with the Adafruit Flora,* Maker Shed item #9781457183225-P at makershed.com. Project help and modeling from Risa Rose.

CIRCUIT DIAGRAM

Each component connects to the Flora main board as follows:

GPS:
3.3V → 3.3V
TX → Flora RX
RX → Flora TX
GND → GND

Accelerometer:
GND → GND
SCL → SCL
SDA → SDA
3.3V → 3.3V

NeoPixel ring:
Vcc (power) → Flora VBATT
IN (Data Input) → Flora D6
Gnd (ground) → GND

Tactile switch:
Any 2 diagonal pins → Flora D10 and GND

Battery:
Plugs in via the onboard JST port

John de Cristofaro for Adafruit

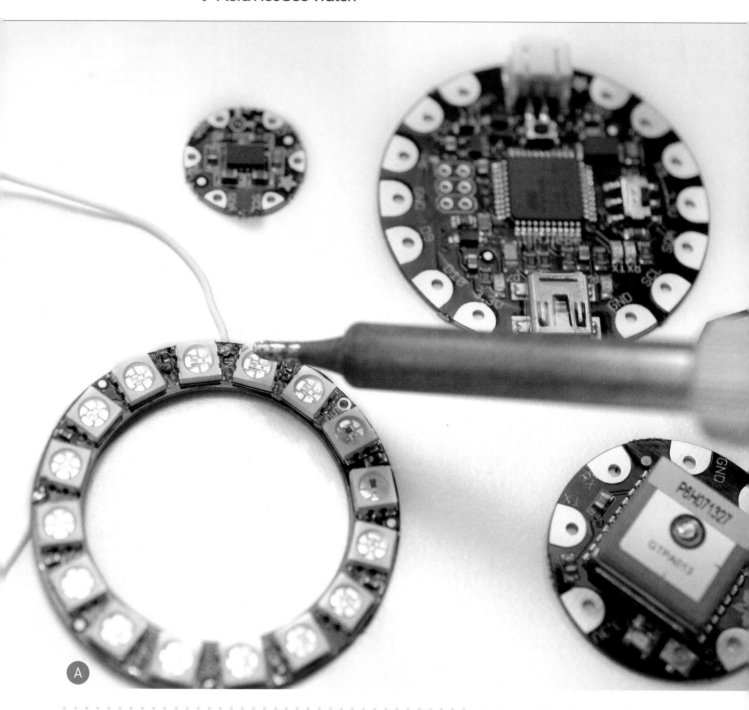

A

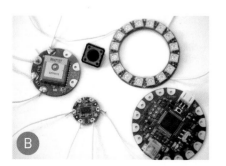

B

C

1. Assemble the circuit

1a. Prepare the components

Start by soldering small stranded wires to your electronics components, about 2" long each. Strip the wire ends and twirl the stranded core to make it pass easily through the circuit boards' holes. Solder wires to the NeoPixel ring's IN, Vcc, and Gnd pads on the back of the board (**Figure A**).

Also solder wires to the GPS module (3.3V, RX, TX, and GND) and the accelerometer/compass module (3V, SDA, SCL, and GND) (**Figure B**).

Trim off any wire ends with flush snips.

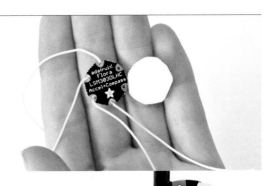

TIP: PUT HEAT-SHRINK TUBING OVER THE GRIPPY ALLIGATOR CLIPS ON YOUR THIRD-HAND TOOL TO PREVENT SCRATCH-ING YOUR CIRCUITS AND COMPONENTS!

1b. Mount the GPS and accelerometer

Use double-stick tape or E6000 glue to mount the GPS in the center back of Flora. Wire the GPS connections to Flora according to the circuit diagram (page 85), keeping the wires short and flush to the board (**Figure C**).

Flip the circuit over and solder the wires to the Flora main board (**Figure D**). Trim any excess wire ends.

Use double-sided foam tape to mount the accelerometer in the center of the Flora, on top of the 32u4 microcontroller chip. The foam helps distance the boards from one another to avoid short circuits (**Figure E**).

Trim, strip, and solder the wire connections for the accelerometer/compass (**Figure F**).

1c. Mount the switch

To prepare the tactile switch, flatten and snip off any 2 diagonal pins (**Figure G**).

Insert the switch into D10 and GND on the component side of the Flora board (**Figure H**). Bend out the leads to hold it in position and solder the joints. This big tactile button makes it easy to switch watch modes by holding down the whole face of the watch for a few seconds.

1d. Test your work

Use a multimeter to verify that your solder connections are secure.

Before you proceed, test the GPS and the accelerometer with the sample sketches provided in their respective code libraries.

1e. Mount the NeoPixel ring

Trim, strip, and solder the NeoPixel ring's wires to the Flora according to the circuit diagram, routing them inside the ring. Load the NeoPixel test code to be sure the ring is connected and functioning properly.

Glue the NeoPixel to Flora, lining up the PCB edges exactly. Don't

NOTE: THE WATCH CODE WILL ALLOW YOU TO ADJUST WHICH LED IS AT 12 O'CLOCK, SO THE ORIENTATION OF THE RING DOESN'T MATTER.

pinch the boards together too much — there should be a cushion of glue between (**Figure I**).

Allow the glue to set for at least 1 hour. The circuit is finished!

2. Assemble the watch

The circuit is held in place by the small strap on a leather watch cuff. The USB and JST ports line up perpendicular to the band for easy access, and the switch is on the "top" of the watch.

Unthread the small strap from one half of the cuff and lay it over the component side of the Flora board. Then glue 2 small strips of fabric onto the circuit board to make "belt loops" (**Figure J**). Clamp with binder clips until dry.

Glide your watch face toward the buckle side and thread the free end back through the cuff (**Figure K**).

3. Upload the code

Grab the NeoGeo watch sketch from makezine.com/neogeo and upload it to Flora. The Flora NeoGeo Watch is very easy to use. The watch fetches the time of day from GPS satellites, so when it first powers on, it needs to get a GPS fix by directly seeing the sky. Once set, the watch automatically displays the current time, with one pixel lit for the hour and another for the minute (**Figure L,** following page).

To change modes after calibration, you'll press and hold the tactile switch near the top of the watch. Compass mode lights up blue in the direction of north (**Figure M** and **Figure N**), no matter which way you turn.

GPS navigation mode points toward the coordinates you configure in the Arduino sketch (**Figure O**), and grows redder the closer you get to your destination.

The code for the Flora NeoGeo Watch is straightforward. We're using the standard Adafruit GPS Library, Time Library, Pololu's

LSM303 accelerometer/compass library, and the Adafruit NeoPixel Library. You'll find links to all required libraries on the NeoGeo Watch Github page. Follow the NeoPixel tutorial to install the library and run the *strandtest.ino* sample code.

Then follow the Flora GPS tutorial to test your GPS module. Make sure your GPS has a direct view of the sky.

Next, calibrate the compass module. Follow the steps on the Pololu LSM303 Github page on how to use the calibration sketch (github. com/pololu/LSM303). Then take the numbers from the calibration sketch and dump them into the NeoGeo Watch sketch in the **Calibration values** section of the code.

In the **WAYPOINTS** section of the code, dump in a location, such as your home, so you'll always be able to find your way home. We like iTouchMap (itouchmap.com) for finding latitudes and longitudes online.

Now you'll calibrate your watch. First calibrate your NeoPixel Ring. Simply light up each pixel using the NeoPixel code to determine which one is your **TOP_LED** pixel (from 0–15).

Then make sure your watch knows which way north is. Select the compass mode. Then, using a compass on your smartphone, or the old-fashioned kind, point your **TOP_LED** north. Count clockwise how many pixels away the lit LED is away from the **TOP_LED**. So, if you aim your **TOP_LED** north, and the LED 4 spots over is lit up, you would change the 0 in the **LED_OFFSET** code above and replace it with 4.

That's it. Upload the code to your Flora, and start using your NeoGeo Watch!

0

NOTE: IT CAN TAKE SEVERAL MINUTES TO ACQUIRE A GPS FIX, BUT IT ONLY NEEDS TO DO THIS ONCE ON POWER-UP. AN OPTIONAL BACKUP BATTERY WILL ALLOW THE GPS TO KEEP ITS FIX AT ALL TIMES.

CAUTION: THIS WATCH IS NOT WATER-PROOF! TAKE IT OFF AND POWER IT DOWN IF YOU GET STUCK IN THE RAIN. DON'T WEAR IT WHILE DOING THE DISHES, ETC.

4. Wear it

Plug in a tiny LiPo battery and tuck it into one of the "side pockets" where the cuff overlaps the strap. A bit of gaffer's tape holds it nicely.

Flora's onboard power switch is hidden under the part of the watch closest to you. Use a fingernail or other pointy, nonconductive object to flip the switch.

Place the watch on a windowsill or go for a walk outdoors to get a GPS fix — the clock will then set automatically, and you'll be ready to navigate through space and time. Enjoy your new Neo Geo Watch! ⊘

See more build photos and share your ideas at makezine.com/flora-neogeo-watch
Share it: #makeprojects

MODES:

WATCH MODE — Shows the "hour hand" as an orange LED, the "minute hand" as a yellow LED, and if both hands are on top of each other, the LED will glow purple. Once the GPS locks on, it will automatically update the time for you. If you lose GPS signal, it will remember the time.

NAVIGATION MODE — Points an LED in the direction you need to go to reach the coordinates you entered in the sketch. When you get close to your destination, the LED will change from yellow to red. Requires a constant GPS lock.

COMPASS MODE — A blue LED always points to the north.

To change between modes, press and hold the button for 2–3 seconds. Hold down the button longer and it will cycle through all 3 modes.

John de Cristofaro for Adafruit

Jeffrey Braverman

Build the Smooth Moods Machine

Soothe ragged nerves with this Arduino-based synthesized sound system.

Written by Gordon McComb

Time Required: An Afternoon
Cost: $80–$100

GORDON MCCOMB

has been building robots since the 1970s and wrote the bestselling *Robot Builder's Bonanza.* You can read his plans to take over the world with an army of mind-controlled automatons, along with other musings, at robotoid.com.

AT NIGHT YOUR SLEEP IS INTERRUPTED BY THE MOURNFUL HOWL OF THE NEIGHBOR'S CAT and the steady drumbeat of a dripping faucet. At work you are greeted by the deafening crescendo of office chatter all around.

Cope with this cacophony of life with the Smooth Moods Machine, a compact and self-contained, remote-controlled music generator built around an Arduino. Programming in the Arduino produces 9 types of relaxing sound and music effects, including wind chimes, surf, and a floating cascade of musical chords.

Inside the Smooth Moods Machine

The basic build is an Arduino Uno, SparkFun's MIDI Musical Instrument Shield, an infrared remote control module, and a couple of wires. Construction takes about 15 minutes — mostly soldering a set of 4 headers to the MIDI shield. An enhanced version uses a homemade interface board to connect the MIDI shield to most any external sound amplifier.

At the heart of Smooth Moods is a tiny MIDI synthesizer, in the form of a surface-mount IC. This chip is conveniently soldered to a shield, making it easy to connect the synthesizer to the Arduino. The Arduino serves as a MIDI controller, the part that literally controls the synthesizer, telling it what to

do. You command the Arduino with an ordinary TV remote control.

You may already be familiar with the concept of MIDI, but here are the essentials: MIDI stands for "musical instrument digital interface," a standard method of producing electronic music. Rather than recording actual sounds, MIDI songs are stored as data, akin to the paper rolls of a player piano. When reproducing music, a MIDI synthesizer mimics different instruments — a grand piano here, an electric guitar there.

Smooth Moods works in a similar way, except instead of playing back a song contained in a file, Smooth Moods generates its own MIDI data on the fly. Instruments and notes are combined at random to give an ethereal quality to the music. Make that music* — with an asterisk. Smooth Moods doesn't play tunes, though it does produce random sequences of chimes, chords, and other sounds that can be considered musical.

So what are you waiting for? Build and program a Smooth Moods Machine, and let your neighbor's cat bother someone else! ●

For the full build, plus materials and tools needed, go to makezine.com/smooth-moods

Share it: *#makeprojects*

PROJECTS

Springboard

Written by Casey Shea ■ Illustrations by Julie West ■ Photo by Gunther Kirsch

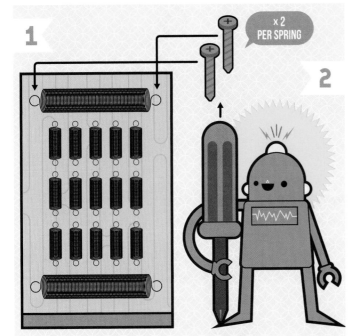

1

x 2 PER SPRING

2

3

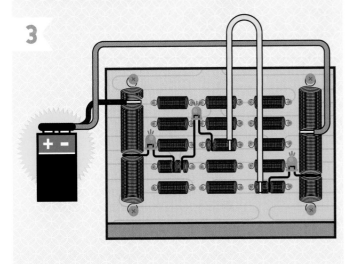

Tell us about your springboard (and show us photos): makezine.com/springboard

Share it: *#makeprojects*

I GET A VICARIOUS THRILL watching young makers' eyes light up like the LEDs in the conductive dough they're working with. But that feeling is too often followed by shared frustration at the increased difficulty when they switch to a solderless breadboard.

To ease the transition, I call back to duty the springboard, a 50-year-old relic that's difficult to find these days, but is easy and cheap to make. The layout mimics a breadboard, so by the time students reach the limitations of the springboard, they're ready to move onto a breadboard.

1. Design your board

Lay out the springs (while not under tension) and mark where you'll attach them to the board. The number of springs depends on the complexity of the circuits you'll build. The most basic design has two vertical 2" springs on the left and right sides, with two rows of ½" springs running horizontally between them, two or three springs in each row. More complicated circuits require more rows. Space the springs so the legs of an LED can easily span two of them.

2. Screw down the springs

Screw them down with your power drill, making sure that none of the screws are touching each other (finish nails are best for boards with closer spacing).

3. Energize

Spread one of the 2" springs with a pointed object (like a multimeter probe) and insert the positive wire from the snap connector. Connect the negative wire to the other 2" spring. Complete the circuit by connecting individual components from spring to spring, using jumper wires when needed. Connect the battery to the snap connector.

The springboard expresses the connected circuit visually and makes it much easier to practice using a multimeter at various locations. ⊘

CASEY SHEA teaches math and Project Make at Analy High School in Sebastopol, Calif. — the hometown of MAKE. In addition to teaching students the skills of making, he is interested in sharing with educators the ways modern tools can be used to create custom instructional materials for their classrooms.

You will need:

» **Plywood or soft wood,** 4"×6"×¾"
» **Springs, 2",** with loops on ends (2)
» **Springs, ½",** with loops on ends (4–15)
» **Wood screws,** ¾" (2 per spring)
» **Power drill**
» **Battery** I used a 9V
» **Snap battery connector, 9V**

WILLIAM GURSTELLE
is a contributing editor
of MAKE. The new and
expanded edition of his book
Backyard Ballistics is available
at all good bookstores.

Time Required:
1 Hour + 1 Week
Cost:
$10

Materials

» **Steel balls, 2" diameter (2)** such as trailer hitches
» **Aluminum foil, heavy-duty**
» **Hydrogen peroxide, 3% solution (optional)**
» **Hydrochloric acid (optional)** often sold as muriatic acid in hardware stores

Tools

» **Orbital sander, sandpaper, or angle grinder**
» **Gloves**
» **Brushes, disposable (optional)**
» **Rag (optional)**
» **Heat gun (optional)**

Hans Goldschmidt and the Invention of Thermite

Re-create the incendiary reaction that welded the world's modern railways. Written by William Gurstelle

"A BLAST FURNACE THAT FITS IN A VEST POCKET."

That's what Wilhelm Ostwald, a Nobel Prize-winning chemist, called thermite, which is a mix of two common chemicals — iron oxide, better known as rust, and powdered aluminum. When combined and ignited, the stuff burns hot enough to melt iron. One of the most powerful chemical processes used in industry, the thermite reaction had a major role in the building of railroads.

Early railroads were built by joining tracks using steel connectors, called fishplates, and several thick metal bolts. However, mechanically joined tracks made an irritating "clack-clack" sound as the trains rode over them. More importantly, bolted connections become loose, requiring a great deal of expensive maintenance.

In 1893, German chemist Dr. Hans Goldschmidt accidentally developed the process for welding thick sections of steel together in the field. While searching for a method of purifying metal ores in his laboratory in Berlin, he discovered that a mixture of iron oxide and aluminum would burn at 3,000°C — more than hot enough to weld steel track. He quickly switched his attention to refining this process, which he named thermite welding. It was first used to weld streetcar track in Essen, Germany, and by the 1930s it was being used nearly everywhere there were tracks to be joined, ushering in the modern practice of continuously welded rail.

Thermite is still used to repair track and in some cases, build new track (**Figure A;** video at makezine.com/thermite-welding). Additionally, the military uses thermite in the form of hand

Elektro-Thermit & Co. KG

grenades to disable captured weapons, such as artillery pieces and truck engines.

THE CHEMISTRY OF THERMITE

One aspect that makes thermite so interesting is the incredible simplicity of the chemical reaction:

$$Fe_2O_3 + 2 Al \rightarrow$$
$$2 Fe + Al_2O_3 + \text{lots of heat}$$

Iron oxide and powdered aluminum are stable at room temperature, and even if you mix them together they'll just sit there as a mound of inert gray powder. But if you can initiate the chemical reaction with an extremely hot flame (hotter than a propane torch) the stuff burns wildly, in an intense exothermic reaction. Oxygen atoms are ripped out of the iron oxide, which then becomes pure metallic iron. It takes a great deal of energy to accomplish this as the oxygen is bound tightly within the rust molecules. The energy comes from aluminum, a powerful reducing agent that heats the iron far past its melting point — enough to spew out sprays of sparks and melt adjacent hunks of iron.

PLAYING WITH THERMITE

Making thermite is not normally a project suitable for amateurs, but this project is safe enough for a junior-high science student to undertake with supervision.

1. Prep the iron

Clamp the threaded portion of the hitch ball into a vise. Using an angle grinder (faster) or sandpaper (slower), remove all of the rust-resistant plating. The ball surface doesn't have to be particularly smooth.

2. Bring on the rust

Immerse the balls in salt water for several days (**Figure 2a**), and allow them to air dry in a dark, warm spot.

OPTIONAL: You can try to jump-start the oxidation process using chemical means. Don rubber gloves and eye protection, then paint a thin coat of acid on the iron (**Figure 2b**).

Let the acid evaporate, then apply the

CAUTION: ⚡ ANGLE GRINDERS THROW OUT A LOT OF HOT METAL, SO WEAR EYE PROTECTION, LEATHER GLOVES, AND A FACE SHIELD.

NOTE: I ATTEMPTED SEVERAL METHODS FOR SPEEDING UP THE RUST-ING PROCESS (ELEC-TROLYSIS, OXYGEN-RICH ENVIRONMENTS, AND SO ON), BUT IN THE END, THE MOST SUCCESSFUL METHOD WAS SIMPLY WAITING A WEEK FOR THE RUST TO BUILD UP.

hydrogen peroxide solution with a brush or rag (**Figure 2c**). The iron will begin to oxidize almost immediately, but this coat of rust is too thin to enable the reaction. Continue to apply the hydrogen peroxide. A hot air gun can help speed the process — just don't touch the metal until it has cooled.

3. Apply the foil

When the balls are thoroughly coated in rust, wrap one in heavy-duty aluminum foil, taking care to make it as smooth and wrinkle-free as possible. (The MAKE Labs had best results with the shiny side out.)

4. Create the reaction

Don heavy gloves. Grasp the threaded rod of the rusted ball in one hand, the aluminum-wrapped ball in the other.

Carefully, but forcefully, bring the rusty ball down smartly on the aluminum-wrapped ball. Watch your fingers! Glancing blows produce extraordinary sparks and a loud, firecracker-like snap. If you examine the balls after striking, you'll see that the aluminum foil has actually welded to the iron.

Rotate the rusted ball after each strike to get a fresh surface of iron oxide, and aim for the areas with the thickest coating of rust. Producing sparks may seem a bit tricky at first, but once you get the hang of it, you can put on quite a show, especially if you dim the lights! ⊘

Tell your friends to try it at makezine.com/thermite-reaction
Share it: **#makeprojects**

A

1

2a

2b

2c

3

Gunther Kirsch

4

Toothy Toothbrush Timer

Hack a novelty toy to help get your dentist-recommended brushing every time. Written and photographed by Steve Hoefer

STEVE HOEFER is a creative swashbuckler, freelance writer, and inventor who regularly contributes projects to the pages of MAKE. His inventions have been featured internationally on TV, over the radio, and in print. He lives on his family farm in Iowa.

EXPERTS SUGGEST YOU BRUSH FOR 2 MINUTES TWICE A DAY, but most people don't come close to that. The Toothy Toothbrush Timer will help you get those recommended 120 seconds each time you brush. It's resistant to steam and occasional splashes, and it will stand up to bathroom conditions to help keep your whites pearly.

How it works

When the toothbrush is lifted, a switch mounted under the holder connects the battery to a timer circuit, which powers a servomotor. As the servo spins, it pushes a rod up and down, opening and closing the teeth. When the time is up, the motor stops and the teeth go silent. Returning the toothbrush to the holder resets the timer by disconnecting the battery.

We could have designed this using an Arduino or other microcontroller, but that would be overkill; all you really need is a 555 timer chip.

1. Prepare the teeth

Chattering teeth come in several different designs; you want to remove any built-in mechanism but keep the hinge intact. Carefully separate the jaws by sliding a knife between them and prying. Remove any hinge pins you find. After you have the jaws apart, remove any screws, the wind-up mechanism, and any springs.

2. Prepare the motor

Now you're going to convert the servomotor into a regular gearmotor. First, put a piece of tape around the front panel of the servo to keep it from

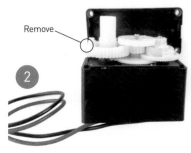

Remove

falling off and spilling gears all over the floor. Then remove the screws and desolder the circuit board from the motor. Remove the board and solder a 6" lead to each motor connection.

Carefully remove the front cover, holding the servo upright so the gears don't fall out. Cut away the nub that prevents the large gear from turning fully, then replace any gears you removed to access it. The motor should turn freely. Put the cover back on and replace the screws. With a fine-bladed saw or a rotary tool, remove the mounting wings from both sides of the case.

3. Assemble the toothbrush holder
Cut the flanged end off the PVC plug. The remaining part should fit smoothly inside a piece of ¾" Class 200 PVC pipe. If not, sand the outside of the plug a bit until it does. Smooth any sharp edges, then wash and clean all the parts. Push a 4" section of pipe into the reducer bushing.

4. Connect the power switch
Cut and strip two 4" wires and solder one to each of the outside terminals on the switch. Using needlenose pliers, carefully bend the tip of the switch lever into a loop.

The "lid" of the project case will actually be the bottom of this project. Measure ¾" in from the center-left edge of the case top and drill three ⁵⁄₃₂" holes to fit the terminals at the bottom of the switch. Thread the wires through these holes and hot-glue the switch in place on top of the case.

5. Wire the electronics
Use a small piece of perfboard to solder the components in place. Be sure to orient the battery holder, 555 timer, electrolytic capacitors, and diode correctly. Install the batteries.

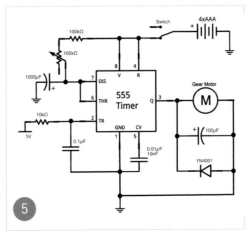

6. Make the teeth chatter
Mount the round servo horn on the servo using the bundled screw. Hot-glue the servo housing inside the case, with the horn toward the bottom. Glue the lower jaw to the top of the case. Drill a ³⁄₃₂" hole through jaw and case, in line with the center of the servo horn. Bend the paper clip into an L, feed it through the hole, and connect it to the horn. Reattach the top jaw. Bend a loop in the clip that just barely touches the inside of the top teeth, then cut off any remaining wire.

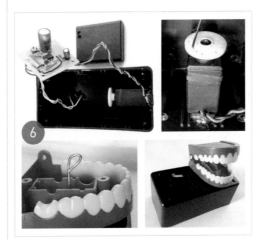

7. Finish up and calibrate
Glue the circuit board to the side of the case and the toothbrush holder over the switch. Slide the plug into the holder, flat side down. Turn on the battery pack and tuck it inside the case. Adjust the trimpot until the motor runs for just over 2 minutes, and seal up the case. Make sure your toothbrush is heavy enough to activate the switch; if not, weight the holder with a stack of nickels.

Use It
Pop a fresh toothbrush in the holder, set it next to the sink, and you're good to go! Smile pretty! ◉

Time Required: A Day
Cost: $35–$45

Materials
» **Chattering teeth novelty toy**
» **Servomotor**
» **Project case, 5"×2½"×2"** RadioShack #270-1803
» **Battery holder, 4×AAA** RadioShack #270-411
» **Switch, SPDT snap action** RadioShack #275-016
» **Proto board, small** RadioShack #276-148
» **555 timer integrated circuit** RadioShack #276-1723
» **Trimpot, 100kΩ**
» **Resistors, 100kΩ and 10kΩ**
» **Electrolytic capacitors, 1000µF and 100µF**
» **Ceramic capacitors, 0.1µF and 0.01µF**
» **Diode,** 1N4001
» **PVC pipe and fittings**
» **Paper clip**

Tools
» **Hot glue gun**
» **Soldering iron**
» **Desoldering braid**
» **Wire cutters and strippers**
» **Needlenose pliers**
» **Screwdrivers**
» **Rotary grinding and cutting tool**
» **Drill with** ⁵⁄₃₂" **and** ³⁄₃₂" **bits**
» **Tape**

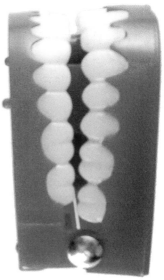

Check out video, photos, and more details at makezine.com/toothy-toothbrush-timer

Share it: **#makeprojects**

GoPro Cannon Cam

3D-print this projectile to take astonishing videos from high above — and look out below.

Written by Dan Spangler

THE ORIGINAL CANNON CAM SHELL WAS ONE OF THE FIRST CUSTOM 3D PROJECTS I MADE when I started working at Make: Labs. Due to the design constraints of the Thing-O-Matic 3D printer, the idea was doomed from the start, but I printed it out anyway. Years later, my editor noticed it and I told him it would probably wobble in flight and shatter on the first crash landing. Intrigued, he asked if I could design a new one that would fly stable and survive a crash. The result was this new 3D-printed GoPro Cannon Cam. When launched from a 3"-bore spud gun, it returns video from the unique perspective of a human cannonball.

A small, lightweight projectile, I reasoned, would survive landings at terminal velocity. After designing an over-engineered rocket 2 feet long and weighing 6 pounds, I pared it down to a smaller body with a shock-absorbing nose cone, camera module, and retractable fins.

1. Print the components

Download the part files at makezine.com/cannon-cam and print them at 50%–100% solid infill for the nose cone impact head, fins, slip rings, and window frame, and 10%–15% infill for the rest (**Figure A**). Line everything up using three 8¾" lengths of 6-32 threaded rod, and make sure all the modules sit flat against one another. Mark the parts with a black Sharpie to ensure the Cannon Cam is always assembled the same way.

2. Assemble the spring-loaded fin module

Cut six ⅝" lengths of ⅛" brass rod, and de-burr the ends. Drill out the holes in all the fins with a #30 bit. Place a torsion spring on a fin so that one of its legs rests in the slot. Trim this leg flush with the fin — about ¼" long — and the other leg to around ¾". Slide the brass rod into the fin (**Figure B**). Insert the ¾" leg of the spring into the small hole on one of the slots on the fin

DAN SPANGLER is the fabricator for Make: Labs, and our in-house dastardly moonlight tinkerer.

Time Required:
1 Week
Cost:
$15–$30

Materials

» **Torsion springs, left-hand, 0.34" OD, 180-degree angle, 0.028" wire diameter (6)** McMaster-Carr part #9271K605, mcmaster.com
» **Brass rod, ⅛"×4"**
» **Threaded rods, #6-32, 8¾" (3)**
» **Allen nuts, #6-32 (3)** McMaster #92066A007
» **Screws, #8-32×½" (7)**
» **Screws, #4-40×⅜" (4)**
» **Clear plastic soda bottle**
» **Compression spring, zinc-plated music wire, 3½" × 0.845" OD × 0.080" wire diameter** McMaster #9657K454

Tools

3D printer
Sharpie, black
Scissors
Hex wrench, 4mm
Drill with #30 bit
Razor
Hacksaw
Screwdriver
File

CAUTION: USE FOR PHOTOGRAPHY PURPOSES ONLY. NEVER AIM YOUR CANNON CAM AT PEOPLE OR PROPERTY.

Gunther Kirsch

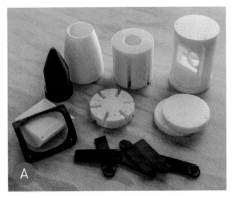

A

B

C

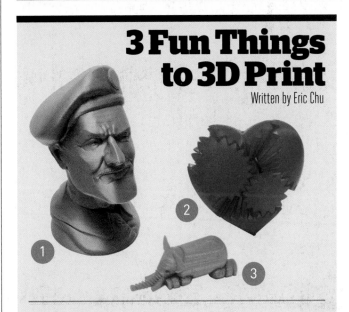

D

module's body cap, then lay the brass pin in the corresponding trough. Repeat with each fin.

Fit the body cap and fin assembly to the fin module body, holding the two parts firmly together, and test all the fins to make sure they don't bind or get stuck in the grooves of the body. If they do, disassemble it and file the sides down until the part no longer binds. Fix the two parts together with three 8-32 × ½" screws.

3. Build the camera module

Carefully peel the label off the soda bottle, then cut off the top and bottom to leave a large clear cylinder. Place your window frame over the plastic and trace the shape of the frame on it, including markings where the screws will enter the body. Cut the window out of the plastic, trimming the corners until it fits in the camera module body. Drill holes to accept the screws.

Trace a line along the window's vertical edges where the all-thread rods will pass through the camera module and cut off the extra material. Place the window in the camera module body and use the 4-40 × ⅜" screws to attach the window frame, sandwiching the window between the two.

4. Final assembly

Put the large spring into its hole in the impact head, then slide both into the bottom of the nose cone body. Screw the 3 all-thread rods into the small holes in the bottom of the nose cone body and add the first slip ring, making sure your tick marks are lined up.

Slide your GoPro Hero3 into the cavity in the camera module, then screw the cover on using 8-32 × ½" screws. Slide the camera module assembly onto the all-thread rods, making sure it's right-way up. Add the second slip ring, followed by the fin assembly. Finally, thread three 6-32 Allen nuts onto the ends of the rods and tighten to hold the whole assembly together. Check that all your marks line up and all your fins spring open freely.

5. Launch It!

Get yourself a spud gun with a 3" barrel. Turn on the GoPro and set it to record video. Tilt the muzzle up to about 80°, and slide the projectile in with the camera facing down. Launch it toward soft, open ground, collect it, and download your footage (**Figure D**). ◑

Download the 3D files and see Cannon Cam aerial videos at
makezine.com/cannon-cam

Share it: *#makeprojects*

3 Fun Things to 3D Print
Written by Eric Chu

1

2

3

1. The Colonel by Ola Sundberg (aka "Sculptor")
thingiverse.com/thing:108867
From the artist who made the Zombie Hunter, The Colonel with his stern smile and distant stare is a high-detail sculpture that prints without the need of support structures.

2. Three Heart Gears by Emmett Lalish
thingiverse.com/thing:243278
Now in 3 different gear ratios, this update of the Screwless Heart Gear uses a new pin design that can be printed in PLA and more reliably on different printers.

3. Elephant by LeFabShop
thingiverse.com/thing:257911
Inspired by the gigantic robotic elephant in Ile de Nantes, France, this cute toy has an articulated head and leg pairs, and even a flexible trunk. Prints as a single print with all parts interlocked.

Toy Inventor's Notebook

BEND A BATTERY BOX

Invented and drawn by Bob Knetzger

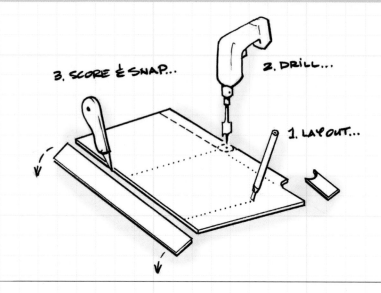

3. SCORE & SNAP...

2. DRILL...

1. LAYOUT...

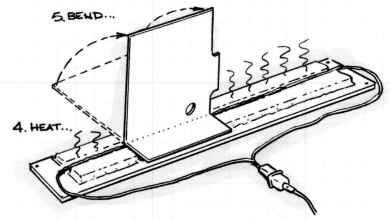

5. BEND...

4. HEAT...

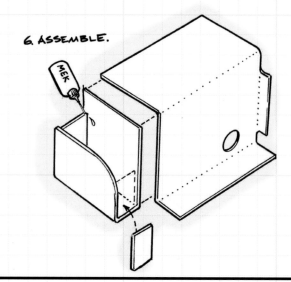

6. ASSEMBLE.

MEK

SOME DAYS THE COOLEST "TOY" IN MY GARAGE/SHOP IS MY SON REED'S ZOOMY LOTUS ELISE SPORTS CAR. On a recent visit, he was frustrated that the original battery cover wouldn't quite fit over his new, larger battery. It was the the perfect opportunity to try out my new strip heater by making a new cover, using a process that works equally well for a producing a new battery box on a kid's Power Wheels jeep.

Strip heaters are great for making accurate bends in thermoplastic parts. I picked up a prewired heater element made by BriskHeat (see MAKE volume 36 for a review). It comes with directions for mounting the element in a plywood channel, lined with heatproof fiberglass and foil. I'd recommend using adhesive-backed foil tape instead; it adheres to the plywood and also traps the fiberglass' fraying edges.

I designed a simple two-part cover, and using a white china marker, laid out the parts on a sheet of 0.09" textured black ABS. I then drilled ½" holes to make radiused inside corners for stress relief, scored the cuts, and bent them to snap apart. To de-burr the sharp edges, I scraped them with the knife. The strip heater softens the plastic evenly and gently for smooth, tight bends. Plan your sequence of bends so that you always have a flat surface to heat.

I bonded the parts together with MEK solvent. A U-shaped bend in the back creates a storage compartment, together with a few side pieces. Self-adhesive velcro spots on the bottom of the S-shaped top flanges hold the removable cover in place inside the Elise's tiny trunk. ◢

To see photos of the finished battery cover, go to makezine.com/bend-a-battery-box

Share it: #makeprojects

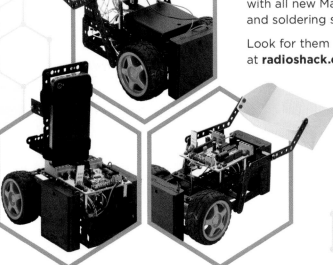

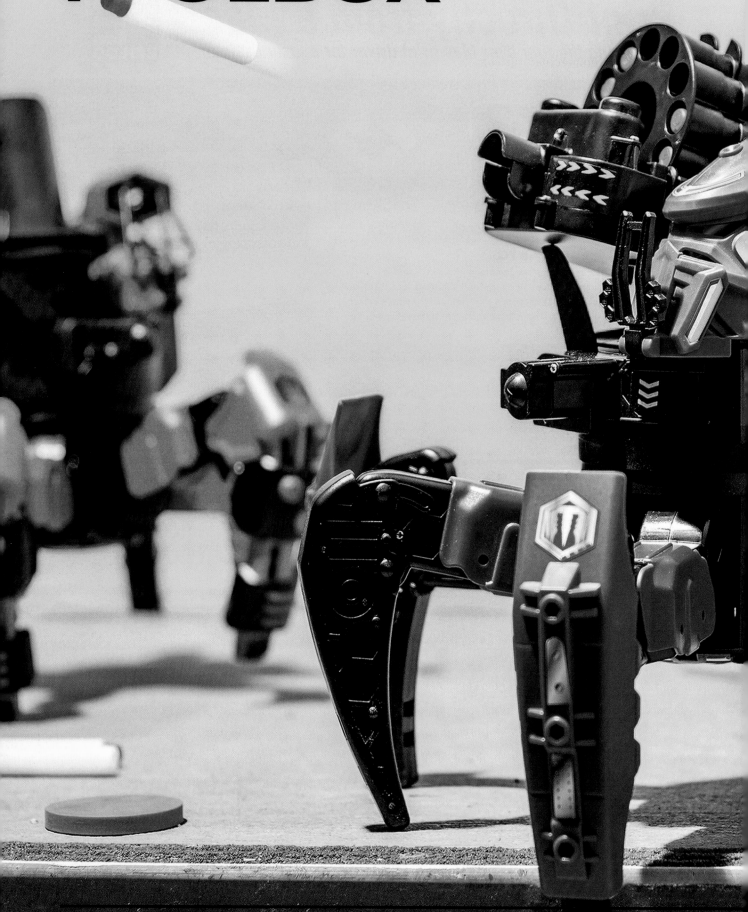

Attacknids Attack!

$80 : combatcreatures.com

Lately, three remote-controlled hexapods have been prowling the halls of MAKE, surprising editors with barrages of soft-tip darts. These Attacknids, from Combat Creatures, are some of the coolest toys in stores today — and better still, they're hackable.

That's thanks in part to Jaimie Mantzel, the innovator, builder, and adventurer who created them. When Mantzel convinced U.K. company Wow! Stuff to carry the toy, he fought to keep the design of the Attacknid open and easy for others to modify. It was an invitation for people to hack the Attacknid, and that's just what they did, like Drake Anthony from styropyro.com, who equipped his Attacknid with a 2-Watt "death ray" laser capable of popping balloons or burning through paper.

Other Attacknid hackers have added autonomous control, ultrasonic range finding, and even a quadcopter that managed to get the rig airborne. MAKE's own Patrick DiJusto took a more peaceful approach on his — he disarmed the Attacknid, replacing its weapons pod with his smartphone, and used Ustream to watch from his computer as he

Thousands of people followed Mantzel's six-year build on his YouTube channel.

Chris Tremblay

remotely controlled the now harmless robot.

Mantzel's design grew out of a giant hexapod robot he built in 2007. He'd been building robots and other cool stuff since he was a kid — including his three-story dome home on a mountain in Vermont — but with this robot he went big, creating a 12-foot-tall by 18-foot-wide crawler.

In the desktop-size version, a unique mechanism actuates all six legs with a single motor, and turns the head with another. The gearing inside the Attacknid makes the robot walk in whatever direction the head is facing. On top of the sheer coolness of the way it looks and moves, the robot has a variety of weapons that shoot foam darts, disks, or pingpong balls. One Attacknid can battle another, blast away its armor, and knock the opponent out with three direct hits to the "Battle Brain."

Hasbro has licensed the design and will be releasing a Nerf-branded version of the Attacknid this fall. And while Hasbro doesn't officially condone mods to their products, there *is* an active Nerf-modding community. We look forward to making our army of six-legged freaks. — *Andrew Terranova*

Gunther Kirsch

Hakko 808 Desolder Tool Kit

$190 : hakko.com

Stop fiddling around with desoldering wicks and those silly blue "sucker" tools. Even if you never make mistakes (or never admit to it), a pro-line desoldering tool like this will change your life for the better, because it makes salvaging components from techno-scrap not just fast and easy but actually pleasurable. And diligent scavenging can recoup the price of the tool pretty quickly. Just touch the nozzle to a PCB pin or solder joint, wait a half a sec for the metal to reflow, and pull the trigger: Zip! The built-in vacuum pump makes a sound like a snoring duck and the solder vanishes up the tube into the catch chamber. It comes in a sturdy plastic case with an accessory kit that includes an extra solder collection chamber, replacement filter elements, nozzle wrench, and miscellaneous cleaning tools. The only thing missing is a benchtop holder, and you'll need a pretty hefty one to support the weight of the gun.
—Sean Michael Ragan

GRIPSTER NUT STARTER

$6 : micromark.com

If you've ever tried to apply rotational force to a small part held with tweezers, then you've probably also spent time on the floor looking for that part. Get off the floor and buy the Gripster Nut Starter. It does a fine job of holding small nuts so they can be threaded onto parts and into hard-to-reach spots. Pushing a plunger on the back end causes four spring steel fingers at the front end to extend and spread. When pressure on the plunger is released, an internal spring causes the fingers to pull in and close, allowing you to hold small objects. I've found it's also great for starting wood and machines screws, as well as for threading tiny washers. It's particularly useful for fishing through containers of small assorted parts and grabbing just the right one. Congratulations, your fingers just got smaller.   —Dug North

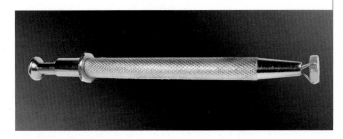

BRICKSTUFF LIGHTING SYSTEM FOR LEGO BRICKS

$55 : brickstuff.com

Brickstuff sells LEDs tiny enough to fit into a small Lego brick. Imagine equipping your Lego starship with a bunch of those lights, with wires thin enough to fit beneath a brick. They also offer an Arduino-based lighting controller. The latter device is a small circuit board with an ATtiny85 microcontroller chip built in. It comes preprogrammed with a dozen lighting schemes that you can cycle through with the touch of a button — no programming needed. Their Lighting Effect Starter Kit comes with the controller, 6 LEDs, connector wires, and a variety of adapter boards.   —John Baichtal

HASEGAWA TOOL MODELING SAW SCRIBER SET

$10 : hlj.com

These modeling saws are by no means new, but I recently discovered this obscure tool line and love it. The photo-etched set of saws is made specifically to scribe styrene, useful for undertakings such as making panel lines on airplane models. These thin saws can also cut thin flashing off of vinyl kits — anything thicker, though, and I reach for my jeweler's saw to be safe.

That said, the numerous curved blades make scribing lines along rounded objects — like airplane fuselages — less laborious. Hasegawa also has flexible, metal scribing templates that are equally worth checking out. —*Jason Babler*

CHARGER DOCTOR

$9 : dx.com

Over the years I have owned various phones and devices that recharge over USB. As a result, I have a collection of USB wallwarts, not all of which charge equally well. While planning to build a meter to figure out which work best, I found the Charger Doctor.

The inline device reads the voltage and current draw of whatever USB gadget is connected. I quickly found out that the USB port on my Mac laptop delivers about 0.4 amps, my Kindle charger delivers about 1 amp, and my Samsung charger delivers 1.7 amps. More surprisingly, one of my 12V car chargers delivers 1.8 amps. When in need of a quick boost, I now know where to go first. —*Greg Brandeau*

DIGITAL SERVO TESTER

$15 : gt-rc.com

The portable, inexpensive tool from GT Power supports up to four servos, powered by a 4.8-6V supply, and two different modes. The automatic mode "sweeps" the servos from one end of their range to the other, while the manual mode allows the user to set the position with the knob, as the display shows the current size of the pulse widths (in milliseconds). Being able to configure the position of servos before hooking them up to a receiver is extremely useful, and I'd recommend this to anyone involved in RC, animatronics, or mechatronics use. —*Eric Weinhoffer*

HIFIBERRY AUDIO CONVERTER

$42 : hifiberry.com

If you find the Raspberry Pi's audio output capabilities lacking, it may be time to invest in a Digital to Analog Converter (DAC). Switzerland-based Modul 9's HifiBerry is just that, providing great quality audio output at a reasonable price point: For around $42, it upgrades your sound capabilities to 24bit, 192kHz audio.

I tested both their DAC and Digi models and was quite impressed at the aural improvement — details are clearer and the audio gains a real sense of space that felt horribly absent when played through the Pi's built-in output.

The DAC version has dual RCA jacks, while the Digi model outputs both coaxial and optical S/PDIF audio. The one downside of the device is it lacks hardware volume control, requiring external audio hardware to be of much use. But as you'll likely be running the HifiBerry through a home stereo amplifier, this mainly only affects headphone listening. —*Wynter Woods*

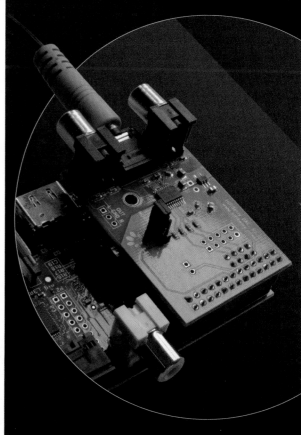

TOOLBOX

SELF-ADJUSTING WIRE STRIPPER

$20 : irwin.com

I'm a fan of quality tools, and these wire strippers will last me for a while. The gripper and stripper segments on the head automatically adjust to quickly strip any type of wire, from 10 gauge to 22 gauge. They also feature a gripper force adjustment knob, cutouts in the handle for crimping connectors, and a wire-cutter, like any good wire stripper should. My only gripe is their inability to reliably strip the soft, rubbery wire I like using for RC applications. Nevertheless, these are always at my side while at the bench. —EW

MARATAC AAA COPPER FLASHLIGHT

$42 : countycomm.com/aaacopper.html

I'll admit it, I covet beautifully crafted, high-quality objects over functionally similar, yet mediocre, mass-produced ones. Call me a tool snob, but I'm enthralled by the Maratac AAA copper flashlight. This single AAA, Cree R3 LED light is lovingly machined from solid copper. The light's driver has two twist-activated brightness modes (low and blinding!), and thanks to the aluminum reflector and coated lens, it throws a very clean spot. I've got plenty of cheap, throwaway lights I've acquired over the years, but this is the one I now clip in my pocket at all times. I look forward to its copper body gaining patina, picking up some history, as I use it to peer into machines, ward off raccoons, and illuminate adventures.

—John Edgar Park

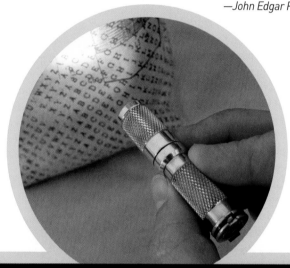

Magic Lantern for Canon EOS cameras

Free : magiclantern.fm

Turn your video-capable DSLR Canon into a high-end cinema camera with Magic Lantern. This free, features-heavy software add-on enables RAW video recording, capturing a much broader spectrum of visual data to your video file and offering a greater latitude for exposure control, color correction, and grading. Additionally, it:

● Improves your camera's audio controls, allowing you to specify input source, set recording levels independently, and record to a separate, higher fidelity .wav file.

● Provides a robust array of focus, exposure, color, composition, and cropping tools through display overlays.

● Offers full control over frame rate, shutter speed, and H.264 bitrate, and, through a bit of a hack (and some post production) makes HDR recordings possible.

Magic Lantern is an open-source product developed by volunteers and released under the GNU public license. Installing it is a breeze, with the software loading from an SD card when the camera is turned on. It is important to note that running it may void your warranty, but if you steer clear of the alpha and nightly releases, it is stable enough for daily, dependable use — many photography professionals swear by the software.

—Tyler Winegarner

The MAKE iPad App is Here!

- Interactive Content
- Video Integration with Tutorials
- Expanded Project Builds

Make:

Find out more at: makezine.com/go/ipadapp

BOOKS

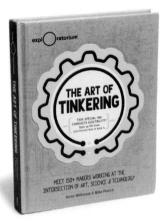

THE ART OF TINKERING

by Karen Wilkinson & Mike Petrich
$33 : Weldon Owen

Not your typical "art" book, *The Art of Tinkering* is not just a must-read, it's a must-make: Like many of the featured tinkerers welcomed to transform the book, you too can hack into the cover's conductive ink to make your own clever circuits. I got lost in *Tinkering*'s pages, awed by its density of information and inspiration. Lushly illustrated profiles pair up with beautifully photographed, relevant projects — each accessible and appealing to makers of any age. Authors Karen Wilkinson and Mike Petrich direct The Tinkering Studio at the Exploratorium in San Francisco, to which they invite people they've discovered at the intersection of art, science, mechanism, and delight. This book takes you there with them. *—Michelle Hlubinka*

LEARN TO PROGRAM WITH SCRATCH

by Majed Marji
$35 : No Starch Press

Create games and art, and learn programming while you're at it. Scratch is a graphical programming language developed at MIT to teach kids how to program and think creatively. By dragging, dropping, and connecting colored blocks, you can create games with animation and sound effects.

The explanations throughout *Learn to Program with Scratch* are excellent and easy to follow. As a STEM enthusiast, I loved author Majed Marji's code examples that used high school-level math for drawing spirograph patterns, like roses and sunflowers. The book also covers the extended capabilities of Scratch 2.0. It's a great resource for schools and public libraries, as well as makerspaces. I know I'll be sharing it with my local programming club.

—Phil Shapiro

21ST CENTURY ROBOT

by Brian David Johnson
$25 : Maker Media, Inc.

I was a little incredulous when I first saw Brian David Johnson's spindly, dare I say adorable, robot of the future. Where are the servos going to go? How are you going to fit all the wires through those tiny little arms and legs? What finally made it click was that "Jimmy" (because every robot has a name) isn't about what we can do now, but what we want to do with the future of robotics. Imagination should drive robotics, not spec sheets.

Johnson is a Futurist-in-Residence for Intel, and this collection of stories uses science fiction to explore the implications of new technologies for the people who will actually use them. Following each story is a "how-to" chapter that describes how the robotic concepts in the story might become a reality. Johnson describes these chapters as unfinished and encourages readers to contribute to the discussion at robots21.com.

—Craig Couden

MAKE: SENSORS

by Tero Karvinen, Kimmo Karvinen, Ville Valtokari
$35 : Maker Media, Inc.

Like a big brother to *Getting Started with Sensors*, *Make: Sensors* takes an in-depth look at using sensors with Arduino and Raspberry Pi platforms to interact with the physical world. Sensors are critical to any kind of interactive electronic project, robot, or environmental monitoring system, and cheap ones are available for sensing all kinds of physical phenomena: humidity, temperature, movement, distance, gases, and light. Even a simple button or volume control knob is a sensor. The trick is understanding how these sensors communicate with the Arduino or Raspberry Pi, and how to write code that not only reads the sensors' values, but accounts for variations in conditions that could cause bad sensor readings. The book's detailed circuit diagrams, illustrations, full-color images, and hands-on projects are easy to parse for beginners, but they're also useful for intermediate users who want to learn more. *—CC*

MAKE: IT ROBOTICS KITS makershed.com /robostart

For the budding roboticist, MAKE and RadioShack have partnered up on an engaging new series of kits that lets you start building bots, and expand them into a variety of fun and clever creations.

The **Make: it Robotics Starter Kit** provides supplies to make a rolling line-follower or walking robot, including motors, sensors, circuit boards and hardware. Add-on kits transform your gadget into anything from a surveillance bot to a mobile mini-catapult, while additional motor kits push the possibilities even further. Simply add an Arduino or other controller and watch young minds get to work.

MAKERSHED
makershed.com

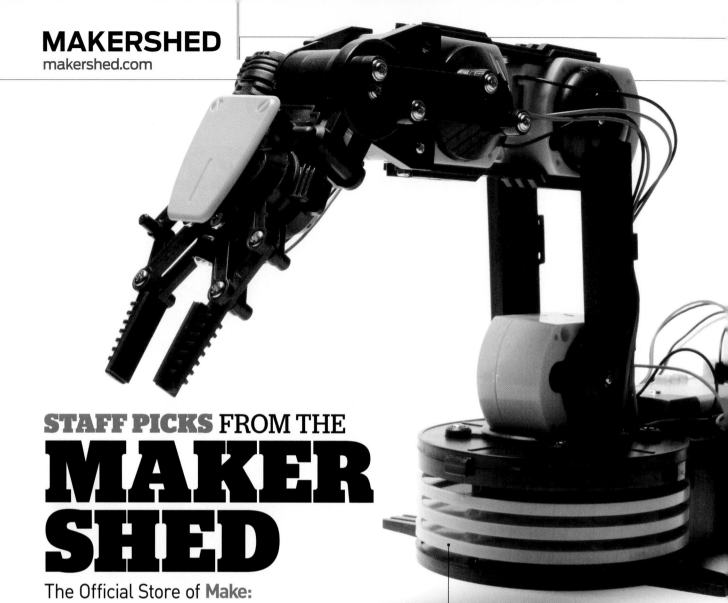

STAFF PICKS FROM THE
MAKER SHED

The Official Store of **Make:**

WE HAVE A SOFT SPOT IN OUR HEARTS FOR ROBOTS because they provide a fun, simple starting point for young makers. Robots promote mechanical thinking and can be created from something as commonplace as a tin can. Even for more advanced tinkerers, robots still hold sway. With the addition of microcontrollers, programming, and engineering skills, robotics kits are as complex or simple as you desire. Whether you're building a bot from an Arduino or a Lego set, you can find supplies and kits to incite creativity at makershed.com.

Written by Natalie Wiersma and Michael Castor

ROBOTIC ARM EDGE
This bot has auto-industry robotic arm functions, but in a compact size. You control the gripper, wrist, elbow, and base joints, and base rotations using the included remote. Want to add intelligence? You can readily hack this arm using an Arduino for automated control.
■ **MKOW07** ■ **$62**

SPARKI
Robotics is one of the most involved areas of making. You need to build a frame, add motors, wheels, accessories, install the electronics, and then worry about programming. Sparki from ArcBotics was designed to make learning robotics easy and fun. Sparki is loaded with features such as a motorized gripper, stepper driven wheels for precise control, a host of onboard sensors, and even a Bluetooth module for wireless communication. The enhanced Arduino IDE includes dozens of sample programs that you can upload and modify so you can spend less time troubleshooting and more time learning.
■ **MKAB2** ■ **$150**

MULTIPLO

Need a completely open-source, robot-focused building system — one with flexible configurations, plug-in sensors, and an easy to use, drag-and-drop IDE? Multiplo delivers, with its plethora of plates, connectors, gears, and other parts allowing for a nearly unlimited number of configurations. Sensors and motors connect directly to the DuinoBot controller and a custom version of the MiniBloq IDE allows for drag-and-drop programming and displays Arduino code in real time to assist learning. Multiplo is perfect for hobbyists, tinkerers, students, and anyone looking to get a start in making and robotics.

■ MKMTP02 ■ $269

CUBELETS

Sometimes you just want to build an interactive robot and not worry about programming. Cubelets are self-contained modules, each with a different function that interacts with the rest once you put them together. They connect together magnetically to form robots with complex behaviors. Each kit includes six Cubelets and a battery charger to get you started. Add more Cubelets for additional functionality and to create your own robot army!

■ MKMR01 ■ $175

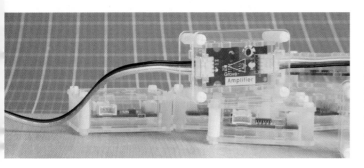

GROVE MIXER PACK

Build functional prototypes quickly and easily with the Grove Mixer Pack's collection of modules. They are colorful, tiny, and accessible enough for young makers and makers new to sensors and microcontrollers. Every kit includes 10 demo cards to get you up and running out of the box. QR Codes on the cards link to a wiki for each project for you to explore and collaborate with others.

■ MKSEEED23 ■ $75

BIT BLOB JR. KIT

Explore funky sounds and digital synthesis with the Bit Blob Jr. Kit from Bleep Labs. Designed to be an introductory soldering kit, the Bit Blob Jr. allows you to create nearly endless sounds and overlays using 8 pads and alligator clip patch cords. Our friends at Bleep Labs call this product "digital noise insanity," and we are apt to agree.

■ MKBL6 ■ $50

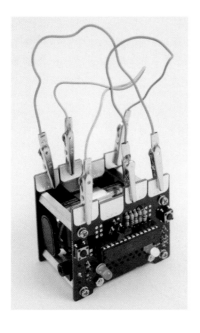

FOLDABLE MINI-SPECTROMETER

Ever wonder what gasses are present in a light bulb or our atmosphere? This Mini-Spectrometer kit is a fun, inexpensive way to get into spectroscopy, the interaction between matter and energy. When excited, matter emits specific wavelengths. Spectrometers allow you to view these wavelengths and determine what a gas is made up of. It includes complete instructions and an attachment to mount it to your phone or webcam for further analysis, calibration, and sharing.

■ MKPBL02 ■ $10

Jared Haworth

My Giant Robotic-Arm Operation Game

Written by Russ Martin

RUSS MARTIN is an inventor and the founder of Florida Robotics. He is the father of two sons (ages 21 and 17), and currently lives in Orlando, Fla., with his wife, Fay, and youngest son, Jesse.

I'VE BEEN CALLED A CROSS BETWEEN JIMMY BUFFETT AND ALBERT EINSTEIN. As a lifelong fan of all types of gadgets mechanical and electronic, I founded Florida Robotics in 1993 with my wife, Fay. Most of our products have been mobile entertainment robots designed to interact with observers, like a robot that moves via voice command and a drink-serving robot that can roll in the sand.

After seeing a medical robot on TV, deciding to build the OpBot — a giant robotic version of the popular board game Operation — was a natural progression for me. In a world where virtually every kid has a sophisticated HD digital game in his pocket, I think many people never get the chance to see and operate an actual mechanical robot with gears whining and motors buzzing. Since OpBot's movements are achieved by mechanical and electrical contacts, it provides an element of realism that

can't quite be duplicated in the virtual word.

Building OpBot was pretty simple, as Florida Robotics routinely uses off-the-shelf components. It only took a few months to complete. The robotic arm is clunky and whimsical, and while outwardly it looks like a bunch of surplus junk assembled by a Rube Goldberg wannabe (it is), the actual working parts are quite reliable. It's precisely controlled by a simple and versatile Basic Stamp 2, and with clever programming, OpBot seems to have a personality that challenges the wits of any avid video gamer.

And since this is a 21st century version of Operation, I decided to update the patient's maladies to fit modern ailments. Forget spare ribs and butterflies in the stomach — OpBot's patient is suffering from a sore texting thumb, iPod ears, and a chip on his shoulder. ●

+ Check out videos of OpBot in action: makezine.com/opbot